Alfred Alcock

Materials for a Carcinological Fauna of India

No. 1: The Brachyura Oxyrhyncha

Alfred Alcock

Materials for a Carcinological Fauna of India
No. 1: The Brachyura Oxyrhyncha

ISBN/EAN: 9783337269838

Printed in Europe, USA, Canada, Australia, Japan

Cover: Foto ©berggeist007 / pixelio.de

More available books at **www.hansebooks.com**

MATERIALS

FOR A

CARCINOLOGICAL FAUNA OF INDIA.

No. 1.

THE BRACHYURA OXYRHYNCHA.

BY

A. ALCOCK, M.B., C.M.Z.S.,

SUPERINTENDENT OF THE INDIAN MUSEUM.

[*Reprinted from the "Journal Asiatic Society of Bengal,"*
Vol. LXIV. Part II, No. 2, 1895.]

Materials for a Carcinological Fauna of India. No. 1. *The Brachyura Oxyrhyncha.*—By A. ALCOCK, M.B., C.M.Z.S., *Superintendent of the Indian Museum.*

Plates III–V.

[Received 11th April:—Read 1st May.]

It was the intention of my immediate predecessor and late friend James Wood-Mason to write a Descriptive Catalogue of the collection of Crustacea in the Indian Museum.

To this end he had collected a very comprehensive Crustacean literature, and had set in motion a scheme for extracting in a handy form the references contained therein.

He had also roughly sorted the whole collection into its component great-groups, and had made a large number of identifications.

In short he had, before his sad and premature death, collected the raw material for, and sketched the broad foundations of, a work that, had he lived on in unimpaired health, might have been a fit companion and sequel to the classical volumes of that great naturalist Henri Milne-Edwards.

Only in the case of the *Stomapoda* had he gone further than this; and I am now preparing to edit, from the rough MS. notes at my disposal, his account of a part of this Order as represented in the collection of the Indian Museum.

The present paper is the first of a series in which I hope to be able to turn to some—though inadequate—account the mass of material accumulated by my predecessor.

My own work in this paper has been to complete, to arrange systematically, to collate, and to verify the available references to the literature of the Oxyrhyncha; to determine about 70 per cent. of the Indian species contained in the collection of the Indian Museum; to prepare the generic diagnoses and the descriptions of all the species mentioned; and to work out, to the best of my ability, keys—which I hope may be of use to naturalists in India—to sub-families, genera, and species.

In the arrangement of the group as a whole, I have been guided and assisted by the *Revision of the Maioid Crustacea,* by Mr. E. J. Miers,

3

in the Journal of the Linnæan Society (Zoology), Vol. XIV. 1879 ; and by the same author's *Report on the ' Challenger' Brachyura* ; and to these important works I have here to acknowledge my great indebtedness.

I have not, however, been able to give my complete adherence to the classification proposed by Mr. Miers, further than to accept the previously adopted division of the Oxyrhyncha into two groups of equal value—the Maioids and the Parthenopoids. To these groups, I would, following Dr. Claus, give the rank of families—*Maiidæ* and *Parthenopidæ.*

But to further sub-divide a group like the Maioids—in which we find, as Miers himself remarks, every reasonable gradation of form from *Stenorhynchus* to *Pericera*—into separate families, as is done by Miers, involves, I think, an unnecessary and unphilosophical interference with the meaning of the term ' family.'

Nor is anything gained, from the point of view of the practical systematist, by establishing families which overlap in all directions.

I am so much indebted to the works of Mr. Miers, that I should be loath to criticize them in any but a friendly spirit. But it seems to me that while Mr. Miers has recognized the value of certain characters round the developments and modifications of which the Maioid Crabs easily cleave into most natural groups, he has proceeded in practice to ignore in great measure the value of his own generalization.

It appears to me that Mr. Miers' families of *Maiinea* consist each of a quite natural nucleus hidden in a loose artificial wrapping.

Beginning with the *Inachidæ* of Miers, we find a natural group, typified by such forms as *Leptopodia* and *Inachus*, linked with forms like *Anamathia, Xenocarcinus, Huenia, Pugettia, Acanthonyx, Doclea* and *Stenocionops*, none of which are any more nearly related to *Leptopodia* and *Inachus* than they are to any other Maioid.

In the *Maiidæ* of Miers again, we find a most arbitrary jumble of forms. Amid the confusion, however, we can discern a large natural nucleus, typified not, it is true, by *Maia*, but by such forms as *Egeria, Chionœcetes, Pisa, Naxia,* etc.; but these are no more nearly related to *Maia, Paramithrax, Schizophrys, Criocarcinus,* and *Micippa* than they are to any other Maioid.

The third family, *Periceridæ,* is even more bewildering; but as Miers himself, in his *Report on the ' Challenger' Brachyura,* has distributed many of his original Periceroid genera among the other two families, it would be unjust to enter into any detailed criticism of this family now.

4

The classification proposed in this paper is in many respects a reversion to the older authors.

For a most interesting and instructive historical and critical review of the Oxyrhyncha as a whole, I would refer to the Introduction of Miers' paper, already cited, in the Journal of the Linnæan Society, Zoology, Vol. XIV. 1879, pp. 634–642.

I have only to add that as almost all the new species described in this paper have been dredged by the 'Investigator,' they will be figured in next year's issue of the "Illustrations of the Zoology of the 'Investigator.'"

Tribe OXYRHYNCHA or MAIOIDEA.

Oxyrinques, Oxyrinchi, Latr. Hist. Nat. Crust. et Insect. tom. VI. p. 85.
Oxyrhinques et *Canceriens Cryptopodes,* Milne-Edwards, Hist. Nat. Crust. tom. I. pp. 263, 368.
Maioidea or *Oxyrhyncha,* Dana, U. S. Expl. Exp. Crust. Pt. I. pp. 66, 67 and 75.
Oxyrhyncha, Miers, Journ. Linn. Soc., Zool., Vol. XIV. 1879, p. 634; and 'Challenger' Brachyura, p. 2.

Carapace more or less narrowed in front, and usually produced to form a rostrum : branchial regions considerably developed, hepatic regions small. Epistome usually large ; buccal cavity quadrate, with the anterior margin usually straight. Branchiæ almost always nine in number on either side *: their efferent channels open at the sides of the endostome or palate. Antennules longitudinally folded. The palp of the external maxillipeds is articulated either at the summit or at the antero-internal angle of the meropodite. The external genitalia of the male are inserted at the bases of the fifth pair of trunk-legs.

The Oxyrhyncha may be sub-divided into two families, namely :—

(1) the *Maiidæ,* in which the basal joint of the antennæ is well developed, and in which it is exceptional to find the chelipeds vastly longer than the other legs ;

and (2) the *Parthenopidæ,* in which the basal joint of the antennæ is very small, and is embedded between the front and the floor of the orbit ; and in which it is exceptional not to find the chelipeds vastly longer and vastly more massive than the other legs.

---

* *Encephaloides* is the only Oxyrhynch known to me in which the branchiæ are less than nine in number on either side : in *Encephaloides* the reduction, both in size and number, of the anterior branchiæ seems to be due to the enormous development of the four posterior branchiæ.

5

Family I. MAIIDÆ.

Macropodiens and *Maïens*, Milne-Edwards, Hist. Nat. Crust. I. 272.
Maiinea, Dana, U. S. Expl. Exp. Crust. Pt. I. pp. 76 and 77, (and *Oncininea*.)
Maiinea, Miers, Journ. Linn. Soc., Zool., Vol. XIV. 1879, p. 640 ; and ' Challenger ' Brachyura, p. 2.

Basal antennal joint well developed, and occupying all the space between the antennulary fossa and the eye.

Taking the characters sagaciously suggested by Miers, namely, the relative development of the component parts of the orbit, including basal antennal joint—-as the basis of a division, the members of the family *Maiidæ* fall into four natural groups or sub-families as follows :—

Key to the Sub-families of Maiidæ.

Sub-family I. *Inachinæ.* Eyes without orbits : the eyestalks, which are generally long, are either non-retractile, or are retractile against the sides of the carapace, or against an acute post-ocular spine that affords no concealment. The basal joint of the antennæ is extremely slender throughout its extent, and is usually long :—

Alliance 1. *Leptopodioida.* Basal joint of the antennæ usually sub-cylindrical, or at any rate convex ventrally, often independent of the neighbouring structures : the external maxillipeds have the merus narrower than the ischium, and the palp large and coarse, and hence have a somewhat pediform appearance.

Alliance 2. *Inachoida.* Basal joint of the antennæ flattened or concave ventrally, and intimately fused with the neighbouring parts ; its antero-external angle often produced to form a spine visible from above : the external maxillipeds have the merus at least as broad as the ischium, and the (small) palp borne at the internal angle of the merus.

Sub-family II. *Acanthonychinæ.* Eyes without true orbits : the eyestalks, which are very short or sometimes even obsolescent, are either concealed beneath a forwardly-produced supra-ocular spine, or are sunk in the sides of a huge beak-like rostrum ; a postocular spine or process is sometimes present, but is not excavated for the reception of the retracted eye. The basal antennal joint is truncate-triangular. The external maxillipeds have the merus as broad as the ischium.

Sub-family III. *Pisinæ.* Eyes with commencing orbits, of which one of the most characteristic parts is a large, blunt, usually but not

6

always isolated, cupped post-ocular process into which the eye is retractile, but never to such an extent as to completely conceal the cornea from dorsal—still less from ventral—view; there is almost always also a distinct supraocular eave, which is sometimes produced forwards as a spine: the eyestalks are short. The basal antennal joint is broad; its antero-external angle is generally produced forwards, as a spine or tooth. The external maxillipeds have the merus as broad as the ischium.

Alliance 1. *Pisoida.* Post-ocular cup distinctly isolated from the supra-ocular eave by a gap or fissure.

Alliance. 2. *Lissoida.* Post-ocular cup in the closest contact with the supra-ocular eave, a suture only intervening.

Sub-family IV. *Maiinæ.* Eyes either (1) with orbits, which may be incomplete or complete, but are always complete enough to entirely conceal the fully retracted cornea from dorsal view; or (2) but partially protected by a huge horn-like or antler-like supra-ocular spine, or by a large jagged post-ocular tooth (*Paramicippa tuburculosa*, Edw.), or by both. The eyestalks are usually long.

The orbit, when present, is formed in one of two ways; there is always an arched—often very strongly arched—supra-ocular eave, and a prominent post-ocular spine; and either (1) the interval between the eave and the spine is filled by another spine, in which case the roof of the orbit, though fissured, is fairly complete; or (2) the supra-ocular eave and the post-ocular spine are in contact with one another above, and below with a process of the basal antennal joint, in which case the orbit has not only a complete or nearly complete roof, but a complete or nearly complete floor also.

The basal antennal joint is always very broad, and is either very extensively produced outwards to aid in forming the floor of the orbit, or is armed distally with one or two large spines.

The external maxillipeds have the merus at least as wide as the ischium.

Alliance 1. *Maioida.* The orbit is formed (1) by a supra-ocular hood, the postero-external angle of which is often produced as a spine, (2) by a sharp post-ocular tooth, and (3) by a spine intercalated between the two. Basal antennal joint broad, but not specially produced to form a floor to the orbit; usually armed at both its anterior angles with a strong spine.

Alliance 2. *Stenocionopoida.* There is no true orbit; but either a huge, outstanding, often more or less hollowed, horn-like or antler-like supra-ocular spine, or a postocular tooth, or both. The basal antennal

7

joint is broad, and either has, or has not, one or both of its anterior angles armed with a strong spine. The merus of the external maxillipeds usually has its antero-external angle strongly dilated ; and the buccal frame is often much wider in front than behind.

Alliance 3. *Periceroida*. The carapace is broadened anteriorly by the outstanding, often tubular, orbits : the orbits are formed (1) by an arched supra-ocular hood, or semi-tubular horn, (2) by a hollowed post-ocular process, and (3) by a remarkable broadening, or by a prolongation, of the anterior part of the basal antennal joint; and they afford complete concealment to the retracted eye. The rostrum is often more or less deflexed.

I am afraid that this last sub-family will, at first, meet with hostile criticism ; but I feel pretty sure that it is a natural group. For, taking the nature of the orbits, eyes, and basal antennal joint as the primary bond of relation, we find, if we exclude the aberrant *Stenocionopoida*, a regular gradation from the imperfect orbit and the narrower basal antennal joint of *Maia*, through the more perfect orbit and broader basal antennal joint of, *e.g.*, *Micippa thalia* and *Micippa cristata*, to the perfect tubular orbit of *Microphrys* (if *Microphrys cornutus* be the type), *Tiarinia* and *Macrocœloma*. The *Stenocionopoida* again are linked on, through *Picrocerus* and *Picroceroides*, to the *Periceroida ;* and, on the other hand, through *Criocarcinus* to the Maioid *Chlorinoides*.

The following is a list of the genera of Maioid Crabs, so far as known to me, arranged in accordance with the afore-proposed classification. Within each sub-family the genera are arranged alphabetically. Indian genera are printed in roman type, and all genera known to me by autopsy are marked with an asterisk.

Complete references are not given ; but only references to the best diagnoses with which I am acquainted. The bibliography of Indian genera will be found in the sequel.

Family **Maiidæ**.

Sub-family I. *Inachinæ*.

ALLIANCE I. LEPTOPODIOIDA. |c₁₁c⁵

* Achæus.
Achæopsis, Stimpson, Proc. Ac. Nat. Sci. Philad., 1857, p. 219.
? *Anisonotus*, A. Milne-Edwards, Miss. Sci. Mex. Crust. I. p. 195.
* Camposcia.

8

Cyrtomaia, Miers, 'Challenger' Brachyura, p. 14.

* Echinoplax.

Ergasticus, A. M-E., Miers, 'Challenger' Brachyura, p. 29.

Ericerus, Mary J. Rathbun, Proc. U. S. Nat. Mus., Vol. XVI. p. 223.

Leptopodia, Leach, Zool. Miscell. II. 15 : Milne-Edwards Hist. Nat. Crust. I. 275 (Synonomy see Miers, Journ. Linn. Soc. Zool. XIV. 1879, p. 643).

Lispognathus, A. Milne-Edwards, Bull. Mus. Comp. Zool. Vol. VIII. 1880-81, p. 9 ; and Miss. Sci. Mex. Crust. I. p. 349 : and Miers ' Challenger' Brachyura, p. 27.

* *Macrocheira*, de Haan, Faun. Japon. Crust., p. 88 : and Miers, ' Challenger ' Brachyura, p. 33.

Metoporaphis, Stimpson, Ann. Lyc. Nat. Hist., New York, Vol. VII. 1862, p. 198.

* Oncinopus.

Pactolus, Leach, Zool. Miscell. II. 19 : Milne-Edwards, Hist. Nat. Crust. II. 189 .

* Paratymolus.

* Platymaia.

Pleistacantha, Miers, P. Z. S., 1879, p. 24.

Podochela, Stimpson, Ann. Lyc. Nat. Hist., New York, Vol. II. 1862, p. 194, (Synon. *Podonema*, Stimpson, Bull. Mus. Comp. Zool., Vol. II. 1870-71, p. 126).

* *Stenorhynchus*, Lamk., Milne-Edwards, Hist. Nat. Crust. I. 278 (Syn. Miers, Journ. Linn. Soc. Zool., XIV. 1879, p. 643).

New genera :—Lambrachæus, Physachæus, Grypachæus.

ALLIANCE II. INACHOIDA.

Anacinetops, Miers, Ann. Mag. Nat. Hist. 1879, Vol. IV. p. 3.

Anasimus, A. Milne-Edwards, Miss. Sci. Mex. Crust. I. p. 360.

Anomalopus, Stimpson, Bull. Mus. Comp. Zool. II. 1870-71, p. 124.

* Apocremnus.

Arachnopsis, Stimpson, Bull. Mus. Comp. Zool. II. 1870-71, p. 121.

Batrachonotus, Stimpson, Bull. Mus. Comp. Zool. II. 1870-71. p. 122.

* Collodes.

* Encephaloides.

Erileptus (? =*Anasimus*), Mary J. Rathbun, Proc. U. S. Nat. Mus. Vol. XVI. 1893, page 226.

? ? ? *Eucinetops*, Stimpson, Ann. Lyc. Nat. Hist. New York, Vol.

9

164   *Carcinological Fauna of India.*

VII. 1862, p. 191 (more probably, as Stimpson himself suggested, allied to *Micippa*).

Euprognatha, Stimpson, Bull. Mus. Comp. Zool. II. 1870-71, p. 122.
Eurypodius, Guérin ; Milne-Edwards, Hist. Nat. Crust. I. 283.
Gonatorhynchus, Haswell, Cat. Austral. Crust., p. 10.
Halimus, Latr., Edw., Milne-Edwards, Hist. Nat. Crust. I. 340.
* *Inachus*, Fabr., Edw., Milne-Edwards, Hist. Nat. Crust. I. 286.
* Inachoides.
* *Microhalimus*, Haswell, Cat. Austral. Crust., p. 7.
Neorhynchus, A. Milne-Edwards, Miss. Sci. Mex. Crust. I. p. 186,
(=*Microrhynchus*, Bell, P. Z. S., 1835, p. 88, and Trans. Z. S. II. 1841,
p. 40).
Oregonia, Dana, U. S. Expl. Exp. Crust. I. p. 105.
Pyromaia, Stimpson, Bull. Mus. Comp. Zool. II. 1870-71, p. 109.
* *Trichoplatus*, A. Milne-Edwards, Ann. Sci. Nat. (6) IV. 1876,
Art. 9, p. 2.

Sub-family, II.   *Acanthonychidæ.*   p. 19

* Acanthonyx.
Antilibinia, Macleay, in Smith's Ill. Zool. S. Africa, p. 56.
Cyclonyx, Miers, Ann. Mag. Nat. Hist., 1879, Vol. IV. p. 6.
Dehaanius, Macleay, in Smith's Ill. Zool. S. Africa, p. 57.
Epialtus, Milne-Edwards, Hist. Nat. Crust. I. 344.
Eupleurodon, Stimpson, Ann. Lyc. Nat. Hist. New York, Vol. X.
1874, p. 98.
Goniothorax, A. Milne-Edwards. Bull. Soc. Philom. (7) III. 1878-79,
p. 103.
* Huenia.
Leucippa, Milne-Edwards, Hist. Nat. Crust. I. 345.
Mimulus, Stimpson, Ann. Lyc. Nat. Hist., New York, Vol. VII.
1860, p. 199.
Peltinia, Dana, U. S. Expl. Exp. Crust. I. p. 129.
* Menæthius.
Mocosoa, Stimpson, Bull. Mus. Comp. Zool. II. 1870-71, p. 128.
* Pugettia.
? * Scyramathia.
* Simocarcinus.
* Sphenocarcinus, (?= *Oxypleurodon*, Meirs, 'Challenger' Brachyura,
p. 38.)
Trigonothir, Miers, Ann. Mag. Nat. Hist. 1879, Vol. IV. p. 4.
* Xenocarcinus.

10

Sub-family III. *Pisinæ.*

.ALLIANCE I. PISOIDA.

Arctopisis, Lamk. see Pisa emend. Miers, *infra.*

Acanthophrys, A. Milne-Edwards (as limited by Miers, J. L. S. Zool. XIV. 656), Ann. Soc. Entom. Fr. (4) V. 1865, p. 141, pl. v. fig. 3.

* *Anamathia,* Roux ; Milne-Edwards, Hist. Nat. Crust. I. 285.

Chionæcetes, Kroyer ; Miers, Journ. Linn. Soc. Zool. XIV. 1879, p. 654 (Syn. *Peloplastus,* see Miers, J. L. S., Zool. XIV. 654).

* *Chorilibinia.*

Chorinus, Leach ; Milne-Edwards, Hist. Nat. Crust. I. 314.

* *Doclea.*

* *Egeria.*

? *Esopus,* A. Milne-Edwards, Miss. Sci. Mex. Crust. I. p. 89.

* *Eurynome,* Leach ; Milne-Edwards, Hist. Nat. Crust. I. 350.

Hoplopisa, A. Milne-Edwards, Bull. Soc. Philom. (7) II. 1877-78, p. 222 ; and Miss. Sci. Mex. Crust. I. p. 201.

* *Hyas,* Leach ; Milne-Edwards, Hist. Nat. Crust. I. 311.

* *Hyastenus* (*Syn.* Lahainia and Chorilia.)

Lepteces, Mary J. Rathbun, P. U. S. N. M., Vol. XVI. 1893, p. 83.

Libidoclea, Edw. and Lucas, Voy. Amer. Merid. Crust., p. 6.

* *Libinia,* Leach ; Milne-Edwards, Hist. Nat. Crust. 1. 298.

Lepidonaxia, Zool. Record, 1877, Crust., p. 11.

Loxorhynchus, Stimpson, Journ. Bost. Soc. Nat. Hist., Vol. VI. 1857, p. 451.

* *Naxia* (*Syn.* Naxioides and Podopisa).

? *Nibilia,* A. Milne-Edwards, Miss. Sci. Mex. Crust. I. p. 132.

Notolopas, Stimpson, Ann. Lyc. Nat. Hist. New York, X. 1874, p. 96.

Pelia Bell, Trans Zool. Soc. II. 1841, p. 44.

* *Pisa,* Leach, Miers ; Miers, ' Challenger ' Brachyura, p. 53.

? *Pisoides,* Edw. and Lucas, Voy. Amer. Merid. Crust., p. 10.

Prionorhynchus, Jacquinot and Lucas, Voy. Pôle Sud, l' Astrolabe et la Zélée, tom. III. Crust., p. 5.

? *Pyria,* Dana, U. S. Expl. Exp. Crust. I. p. 96.

Rachinia, A. Milne-Edwards, Miss. Sci. Mex., pl. xviii., fig. 1 (if this genus is distinct from *Scyramathia*).

Salacia, Edw. and Lucas, Voy. Amer. Merid. Crust., p. 12.

Scyra, Dana, U. S. Expl. Exp. Crust. I. p. 95.

? * *Scyramathia* (*Syn.* ? *Rachinia*).

Trachymaia, A. Milne-Edwards, Bull. Mus. Comp. Zool. VIII. 1880-81, p. 3 ; and Miss. Sci. Mex. Crust. I. p. 351.

11

ALLIANCE II. LISSOIDA.

? *Coelocerus*, A. Milne-Edwards, Miss. Sci. Mex. Crust. I. p. 84.
Herbstia, Milne-Edwards, Hist. Nat. Crust. I. 301 (Syn. *Rhodia*,
Bell, T. Z. S. II. 1841, p. 43 ; *Micropisa*, Stimpson, Proc. Ac. Nat. Sci.
Philad., 1857, p. 217 : *Herbstiella*, Stimpson, Ann. Lyc. Nat. Hist. New
York, X. 1874, p. 93).
* Hoplophrys.
Lissa, Leach; Milne-Edwards, Hist. Nat Crust. I. 310.
Parathoe, Miers, Ann. Mag. Nat. Hist, 1879, Vol. IV. p. 16.
Perinea, Dana, U. S. Expl. Exp. Crust. I. p. 114.
* Tylocarcinus.

Sub-family IV. *Maiinæ.*

ALLIANCE I. MAIOIDA.

* Cyclax (Cyclomaia).
* Maia.
Maiella, Ortmann, Zool. Jahrb. Syst. &c., VII. 1893-94, p. 51.
Maiopsis, Faxon, Bull. Mus. Comp. Zool., XXIV. 1893, p. 150.
Nemausa, A. Milne-Edwards, Miss. Sci. Mex. Crust. I. p. 80.
* Paramithrax (* Leptomithrax, * Chlorinoides).
? *Phycodes*, A. Milne-Edwards, Rev. et Mag Zool. (2) XXI. 1869,
p. 374.
? *Pleurophricus*, A. Milne-Edwards, Journ. Mus. Godeffr., I. Crust.
p. 260.
* Schizophrys (Dione).
Temnonotus, A. Milne-Edwards, Miss. Sic. Mex. Crust. I. p. 82.

ALLIANCE II. STENOCIONOPOIDA.

* Criocarcinus.
? *Eucinetops*, Stimpson, Ann. Lyc. Nat. Hist. New York, VII.
1862, p. 191.
* *Paramicippa*, Edw. Milne-Edwards, Hist. Nat. Crust. I. 332.
Picrocerus, A. Milne-Edwards, Ann. Soc. Ent. Fr. (4) V. 1865, p. 136.
Pseudomicippa, Heller, Crust. Roth. Meer., SB. Ak. Wien, XLIII.
1861, p. 301 ; and Miers ' Challenger' Brachyura, p. 68 (*nec syn.* Micro-
halimus).
Stenocionops.
Stilbognathus, E. Martens, Verh. zool.-bot. Ges. Wien, XVI. 1866,
p. 379.
Tyche, Bell, P. Z. S. 1835, p. 172, and T. Z. S. II. 1841, p. 58 (*syn.*
Platyrinchus, Desbonne and Schramm, Crust. Guadeloupe, p. 3).

12

ALLIANCE III. PERICEROIDA.

? *Ala,* Lockington, Proc. Calif. Acad. Sci. VII. 1876, p. 65.

Anaptychus, Stimpson, Ann. Lyc. Nat. Hist. New York, VII. 1862, p. 183.

? *Coelocerus,* A Milne-Edwards, Miss. Sci. Mex. Crust I. p. 84.

Cyclocoeloma, Miers, Ann. Mag. Nat. Hist. 1880, Vol. V. p. 228.

* Cyphocarcinus.

Hemus, A. Milne-Edwards, Miss. Sci. Mex. Crust. I. p. 88.

Leptopisa, Stimpson, Bull. Mus. Comp. Zool. II. 1870-71, p. 114.

* Macrocoeloma (Entomonyx : both these genera of Miers seem to me to be synonymous with *Micippoides* of A. Milne-Edwards.)

* Micippa.

Micippoides, A. Milne-Edwards, Journ. Mus. Godeffr. I. Crust. 254 (probably *Macroceloma* and *Entomonyx* may be here included).

* *Microphrys,* Edw.; Milne-Edwards, Ann. Sci. Nat. Zool. (3) XVI. 1851, p. 251; and Miers, ' Challenger ' Brachyura, p. 82 (*syn. Milnia,* Stimpson, Ann. Lyc. Nat. Hist. New York, VII. 1862, p. 179 : *Omalacantha,* Hale Streets, Proc. Ac. Nat. Sci. Philad. 1871, p. 238; and A. Milne-Edwards, Miss. Sci. Mex. Crust. I. p. 64 : *Fisheria,* Lockington, Proc. Calif. Ac. Sci, VII. 1876, p. 72.

Mithrax, Leach ; Milne-Edwards, Hist. Nat. Crust. I. 317 ; and Miers, ' Challenger' Brachyura, p. 84 (*syn. Mithraculus,* White, *vide* Miers. J. L. S., Zool. XIV. 1879, p. 667 : *Teleophrys,* Stimpson, Amer. Journ. Sci and Arts. (2) XXIX. 1860, p. 133.)

Othonia, Bell (*Pitho,* Bell, P. Z. S. 1835, p. 172 : *Othonia,* Bell T. Z. S. II. 55) : and A. Milne-Edwards, Miss. Sci. Mex. Crust. I. p. 114.

Pericera, Latr., Edw. ; Milne-Edwards, Hist. Nat. Crust. I. 334 ; and Miers, ' Challenger ' Brachyura, p. 76.

Picroceroides, Miers, ' Challenger' Brachyura, p. 77.

(This genus, though placed in this alliance on account of the structure of the orbits and basal antennal joint, is in many respects more closely allied to the Stenocionopoida).

Sisyphus, Desbonne Schramm, Crust. Guadeloupe, p. 20.

? *Thoe,* Bell, P. Z. S., 1835, p. 171 : A. Milne-Edwards, Miss. Sci. Mex. Crust. I. p. 120 (*syn., sec.* Miers J. L. S. Zool. XIV. 667 ; *Platypes,* Lockington, Proc. Calif. Ac. Sci. VII. 1876, p. 41).

* Tiarinia.

The genus *Podohuenia,* placed among the Periceridæ in the Zoological Record for 1892 (Crust., p. 17), is inaccessible to me. The reference in the Zoological Record is to Boll. Soc. Nat. Napoli, III. 1889, p. 180.

13

Sub-family INACHINÆ (see Table I.).

Alliance I. LEPTOPODIOIDA (see Table I.).

LAMBRACHÆUS, n. gen.

Closely allied to *Leptopodia* and *Metoporaphis*, from which it differs (1) in its extremely long sub-cylindrical neck, (2) in its minute antennæ and (3) in the *Lambrus*-like proportions of its chelipeds.

Eyes antennules and antennæ borne at the end of a long narrow subcylindrical "neck," which is continued onwards as an extremely long slender spiny rostrum.

Eyes stoutish, salient and non-retracticle: no defined orbits: a small postocular spine. Antennæ minute, exposed to dorsal view. Chelipeds stout and extremely long, with long sub-cylindrical palms and short fingers.

Legs very slender: shorter than the chelipeds.

Lambrachæus ramifer, n. sp., Plate III. fig. 1.

The body is formed by (1) a small trunk, (2) a long narrow almost cylindrical prestomial "neck," and (3) a long slender sinuous spiny rostrum shaped like a withered branch.

The carapace proper is trilobed, the lateral lobes being formed by the branchial regions, and the front lobe being formed by the wings of the buccal frame.

The "neck," at the end of which are borne the eyes, antennules, and antennæ, is rather longer than the carapace proper.

The rostrum is nearly twice the combined length of the neck and carapace.

The eyes are salient and non-retractile, and though there is a narrow dorsal eave round the base of the eyestalks and a pair of tiny postocular spines, there is nothing like an orbit present. The cornea is surmounted by a little tooth.

The antennæ are minute and filiform, and are completely exposed: their total length is not one-sixth that of the rostrum.

The antennules are of large proportions: they fold longitudinally, but when folded are much beyond the capacity of the narrow shallow antennulary fossæ.

The external maxillipeds have broad endopodites, and completely cover the buccal frame: the merus is expanded in both directions, but most at its internal angle, so that the flagellum is inserted nearer to the external angle.

ust an acute post-ocular spinule or spine that
afford im either simple, or two-spined, or emarginate
(in *Pl*

face, and independent.   External maxillipeds
with

le and much longer than
....   LAMBRACHÆUS.

d ap-   *a.*⁴ R o s t r u m
very   formed of   two
some   l o n g   spines;
spi-   none of the legs
talks   subchelate........   ECHINOPLAX.
back-
ever   *b.*⁴ R o s t r u m
short, bifid: last
pair of legs sub-
I.   C   chelate.............   GRYPACHÆUS.
t y
rhy d appendages smooth or
elo few spines: no post-ocu-
gu the   eye-stalks hardly
for ackwards......................   ACHÆUS.

hort, not reaching to the
....   PARATYMOLUS.

l antennal joint project-
....   PHYSACHÆUS.

eye-stalks much curved:
....   CAMPOSCIA.

ce: no post-ocular spine:
....   ONCINOPUS.

II.   oncealment: basal anten-
....   PLATYMAIA.

Allia unding parts, its antero-
ee merus as broad as or
h

, and thus concealing, the
adult, many times the
I.   R..........................   ENCEPHALOIDES.
n
al branchial regions: 2nd
....   INACHOIDES.

....   APOCREMNUS.
II.   R
e..................................   COLLODES.

Table I. Sub-family INACHINÆ.

Eyes without orbits; the eye-stalks usually long and slender, and either non-retractile, or retractile against the carapace or against an acute post-ocular spinule or spine that affords no concealment. The basal joint of the antennæ is extremely slender throughout, and is usually long. Legs slender. Rostrum either simple, or two-spined, or emarginate (in *Platymaia* apparently trifid).

Key to the Indian Genera.

Alliance 1. LEPTOPODINÆ. Antennæ with the basal joint usually sub-cylindrical, or at any rate usually convex on the ventral surface, and independent. External maxillipeds with the merus narrower than the ischium, and often with a large coarse palp, and therefore somewhat pediform in shape.

I. Carapace of the typical Oxyrhynch shape, oblong-triangular or pyriform.	1. Carapace well-shielded, not depressed; rostrum separated from the carapace by a distinct post-ocular constriction, which sometimes forms a long "neck."	ii. Chelipeds never approaching the longest legs in length: rostrum bifid and never approaching the carapace in length: post-ocular neck, when distinct, short.	*a.* Free joints of the antennal peduncle slender, cylindrical and not coarsely hirsute: eye-stalks almost straight: rostrum horizontal.	*a.¹* Eye-stalks salient, but freely moveable forwards and towards and towards backwards: basal antennal joint not reaching to the apex of the rostrum.	*a.²* Basal antennal joint long: cylindrical spacious.	*a.³* Body and appendages very spiny: some post-ocular spinules: cye-stalks retractile backwards, but never concealed.	*a.⁴* Rostrum formed of two long spines; none of the legs subchelate..... ECHINOPLAX.
							b.⁴ Rostrum short, bifid: but pair of legs subchelate....... CYPTACHÆUS.
						b.³ Body and appendages smooth or with very few spines: no post-ocular spine: the eye-stalks hardly moveable backwards.............	ACHÆUS.
					b.² Basal antennal joint very short, not reaching to the front: epistome very narrow..............		PARATYMOLUS.
				b.¹ Eye-stalks salient and rigidly immoveable: basal antenna joint projecting beyond and above the apex of the rostrum..........			PHYSACHÆUS.
			b. Free joints of the antennal peduncle short, flat, and densely hairy: eye-stalks much curved: rostrum somewhat depressed: a post-ocular tooth.....				CAMPOSCIA.
		Chelipeds both markedly longer and vastly stouter than the longest legs: rostrum simple and much longer than the carapace: post-ocular "neck" at least as long as the carapace...............					LAMBRACHÆUS.
	2. Carapace semi-membranous, exceedingly depressed and flat: rostrum in unbroken continuity with the carapace: no post-ocular spine: the last pair of legs subdorsal in position...............						ONCINOPUS.
II. Carapace nearly circular. [Epistome narrow: a large post-ocular spine against which the eye is retractile, but which affords no concealment: basal antennal joint perfectly free, legs long, with much flattened blade-like joints: rostrum trifid.].....................							PLATYMAIA.

Alliance 2. INACHOIDA. Antennæ with the basal joint flattened or concave on the ventral surface, and intimately fused with the surrounding parts, its antero-external angle produced to form a spine which is visible from above on either side of the rostrum. External maxillipeds with the merus as broad as or broader than the ischium, and with the palp small.

I. Rostrum simple: post-ocular spine small: basal antennal spine small or moderate.	1. Branchial regions upraised, and meeting across and thus concealing, the cardiac region: 3rd pair of trunk-legs, in the adult, many times the length of the carapace...................	BUCEPHALOIDES.
	2. Cardiac region not encroached upon by the normal branchial regions: 3rd pair of trunk-legs of moderate length........................	INACHOIDES.
II. Rostrum bifid: post-ocular spine large: basal antennal spine large...	1. Eyes hardly retractile.............................	APOCREMNUS.
	2. Eyes retractile against a strong post-ocular spine...........	CULLODES.

The chelipeds, though actually slender, are relatively to the carapace as stout and long as those of the longer-armed species of *Lambrus*: they are one-third longer than the combined carapace neck and rostrum : they are sub-cylindrical and spiny : their proportions are much those of *Lambrus*, the fingers being not much more than a quarter the length of the palm. The fingers are curved, and are in contact only at their tips.

The legs, which are very slender and are not quite so long as the chelipeds, display no remarkable characters.

The figure, which represents a male magnified two diameters, shows the proportions better than any table of measurements.

Loc. Port Blair, Andaman Islands.

ACHÆUS, Leach.

Achæus, Leach, Malac. Podophth. Brit., Tab. XXII. fig. C.
Achæus, Desmarest, Consid. Gen. Crust., p. 153.
Achæus, Milne-Edwards, Hist. Nat. Crust. I. 281.
Achæus, Miers, Journ. Linn. Soc., Zool., Vol. XIV. 1879, p. 643; and 'Challenger' Brachyura, p. 8.

Carapace triangular with the branchial regions swollen, always more or less constricted behind the eyes. Rostrum very short, bifid. Eye-stalks long and hardly retractile backwards : no orbits or post-ocular spine. Antennæ with the basal joint very slender, sub-cylindrical, the other joints and the flagellum completely exposed. External maxillipeds with the meropodite long, narrower than the ischiopodite, and carrying the next joint at, or near, its apex. Chelipeds short, not very stout. Legs slender, sometimes long and filiform : the dactyli of those of the last two pairs more or less falcate. Abdomen consisting of six segments in both sexes.

As Miers has remarked, this genus is distinguished from *Stenorhynchus* only by the form of the rostrum, which consists of two short lobes instead of two long spines.

Key to the Indian species of the genus Achæus.

I. Carapace with a post-ocular constriction, but with no
    long post-ocular " neck : " dactyli of last pair, or two
    pair, of legs strongly falciform :—
    1. Carapace and eye-stalks smooth ... *A. lacertosus.*
    2. Carapace with a bilobed prominence on the cardiac
        region : eye-stalks with a tubercle on the an-
        terior surface :—
        i. Gastric region smooth .:. ... *A. affinis.*

15

    ii. Gastric region with a sharp tubercle or
      spine ...   . *A. spinosus.*
 II. Carapace with a long post-ocular neck: dactyli of
   last pair of legs hardly curved :—
   1. Lobes of rostrum with a spinate carina: median
     tubercles of carapace low and blunt ... *A. cadelli.*
   2. Lobes of rostrum with a smooth carina: median
     tubercles of carapace sharp and elevated *A. tenuicollis.*

Achæus tenuicollis, Miers.

Achæus tenuicollis, Miers 'Challenger' Brachyura, p. 9, Pl. I. fig. 3.

 "The body is thinly clothed with short curled hairs; the limbs with similar hairs, interspersed among which are some longer ones. The carapace is subtriangulate, little longer than broad, with a neck-like constriction behind the orbits, and armed with spines as follows :—Three conical spines upon the gastric and another upon the cardiac region, two shorter conical spines or tubercles whereof the anterior is the smallest, on each branchial region, behind these one very small on the posterior margin of the carapace, and another on the sides of the branchial regions above the bases of the chelipedes; also a small spine upon the rounded, lateral, hepatic protuberance, and another behind this, on the pterygostomian region; there is also a strong spinule on the upper margin of the orbit, above the eye-peduncles. The lobes of the rostrum are short, and terminate each in a spine. The sternal surface of the body bears a few spinules. The post-abdomen of the male, is as usual, six-jointed (the two last joints having coalesced). The eye-peduncles are robust, with the corneæ protuberant; a small spinule exists on the inferior margin of the eye-peduncle, and another on the upper margin of the eye, near the distal extremity. The antennules are lodged in deep longitudinal fossettes; the very slender basal joint of the antennæ is joined with the front at its distal extremity and bears several small spinules on its inferior surface, the following joint is short, the next about as long as the basal joint, flagella slender; the ischium-joint of the outer maxillipedes is produced at its inner and distal angle which is rounded and bears several spinules on its outer surface, as does also the merus-joint which is rounded, not truncated, at the distal extremity where it bears the next joint. The chelipedes (in the male) are rather slender, and longer than the body; with the joints clothed with rather long hairs; ischium and merus-joints with a series of spinules on their antero- and postero-inferior faces, wrist about as long as palm, with a few spinules hardly discernible amid the hairs which clothe this joint,

16

palm slightly compressed, not dilated, armed with spinules on its upper
and lower margins, fingers about as long as palm, and slightly incurved
at the apices which are nearly destitute of hair; the ambulatory legs
are very slender and elongated; the dactyli of the first three pairs are
short and nearly straight, in the last pair only are they slightly falciform.
Colour (in spirit) light yellowish-brown." (Miers).

A single specimen is included in the Museum collection: the locali-
ty is not quite certain, but it came most probably from the Andamans.

Achæus cadelli, n. sp. Plate V. fig. 1.

In general form and proportions much resembling *Achæus lorina*
(Ad. & White), from which it differs in having the legs even more
slender, and the eye-stalks quite smooth.

The regions of the pyriform carapace are well demarcated, the
hepatic regions being each produced to form a strong sharp tooth.
There are three elevations, arranged in triangle, on the gastric region,
and two, side by side, on the cardiac region.

The rostrum has the usual *Achæus*-form, but each lobe is dorsally
carinate, the carina being spinate or serrate.

Behind the rostrum is a long constricted "neck," more pronounced
even than that of *A. tenuicollis* and *brevirostris.*

The chelipeds are of the usual form. The legs are extremely long
and slender, those of the second trunk segment being about five times
the length of the carapace, rostrum included. The dactyli of the 4th
and 5th pairs are hardly falciform. Length of carapace, 7 millim :
greatest breadth of carapace, 4 millim. : length of 2nd pair of trunk-legs,
36·5 millim.

Loc. Andamans.

Achæus spinosus, Miers.

Achæus spinosus, Miers, Japanese and Corean Crustacea, in Proc. Zool. Soc.,
1879, p. 25.

Carapace triangular, narrowed behind the eyes, and armed with
six spines above, namely : one on the gastric, one — bilobed—on the
cardiac, and two on each branchial region : there are also some spines
or sharp tubercles on the ventrad aspect of the hepatic and branchial
regions. The rostrum is small and bilobed. The eye-stalks are robust,
and have a strong tubercle near the middle of the anterior surface.
Chelipeds in the male robust, the arm and wrist granular above, the
palm swollen, with about six spinules on the upper margin and a few
granules on the lower margin near its base : fingers, in the male, acute

17

with a wide hiatus at base when closed, both with a strong tooth on their opposed margins near the base, and with the outer margins carinate. In the female the chelipeds differ only in being much less robust, and in having the fingers much more closely apposable and toothless. Ambulatory legs long and slender : the dactylus of the last pair strongly falcate.

[The basal antennal joint has one or two spines at its distal end, and the free portion of the antenna is much shorter than the carapace.]

Length of adult, 6 to 7 millim.

In the Museum collection, from the Persian Gulf. Ex coll. W. T. Blanford.

Achæus lacertosus, Stimpson.

Achæus lacertosus, Stimpson, Proc. Acad. Nat. Sci. Philad., 1857, p. 218.

Achæus breviceps, Haswell, Proc. Linn. Soc., N. S. Wales, Vol. IV. 1879, p. 433 (*sec.* Haswell).

Achæus lacertosus and *breviceps,* Haswell, Cat. Austr. Stalk and Sess. eyed Crust., p. 3.

Achæus lacertosus Miers, Zool. "Alert," pp. 181 and 188 ; and "Challenger" Brachyura, p. 8.

Achæus lacertosus, J. R. Henderson, Trans Linn. Soc., Zool., 1893, p. 341.

Carapace triangular, with the regions fairly well delimited and the surface quite smooth beneath a slight pubescence : hepatic region with a horizontal laminar tooth. Rostrum as long as wide, bilobed. Antennæ filiform, the free portion longer than the carapace. Eye-stalks long, slender, smooth. Chelipeds much stouter than the other legs, the meropodite being the stoutest joint, and the hand being incurved and the fingers compressed. The ambulatory legs are long and slender, the first pair being more than three times the length of the carapace : the dactyli of the last two pairs are strongly falcate.

Length of adult about 6 millim.

In the Museum collection are numerous specimens from the Andamans, from Palk Straits, and from the Orissa Coast.

Achæus affinis, Miers.

Achæus affinis, Miers, Zoology of the 'Alert,' pp. 181 and 188, and "Challenger" Brachyura, p. 8.

Achæus affinis, de Man, Archiv. f. Naturges., LIII. 1887, p. 218.

Achæus affinis, Henderson, Trans. Linn. Soc., Zool. (2) V. 1893, p. 341.

Achæus affinis, Ortmann, Zool. Forsch. in Austr. and Malay Arch., Jena, 1894, p. 37.

"Carapace subtriangular and moderately convex, with the surface uneven, but the regions not very distinctly defined ; the post-orbital

18

region is constricted. The rostrum is moderately prominent, the frontal lobes very small and subacute. On the cardiac region is a bilobated prominence, which is usually very much elevated; there is a small angulated prominence on the hepatic regions, and occasionally one or two granules on the branchial regions, which are not at all convex. Eye-peduncles with a blunt tubercle in the middle of their anterior margins. The merus-joints of the outer maxillipedes are narrowed and subacute at their distal ends, where they are articulated with the next joints. The chelipedes (in both sexes) are rather slender; margins of the arm, wrist, and palm usually with a few granules or spinules; merus somewhat trigonous; fingers as long as the palm, and somewhat incurved, with their inner margins denticulated, and having between them when closed (in the males) a small hiatus at base. The ambulatory legs are slender, filiform, and very much elongated, the second legs being, in an adult male, four times as long as the postfrontal por-. tion of the carapace; the dactyli of the two posterior pairs only are distinctly falciform; both chelipedes and ambulatory legs are scantily clothed with long hairs. Length of carapace (including rostrum) of an adult male about 5 lines (10·5 millim.), breadth about 3 lines (6 millim.); length of second leg about 1 inch 8 lines (42 millim.); an adult female has the carapace relatively somewhat broader, length nearly 5½ lines (12 millim.), breadth 4 lines (8·5 millim.).

The bilobated prominence on the cardiac region and tuberculated eye-peduncles serve to distinguish this species." (Miers).

This species is included in the Indian Fauna on the authority of Professor Henderson: there are no specimens in the Indian Museum collection.

PARATYMOLUS, Miers.

Paratymolus, Miers, P. Z. S., 1879, p. 45.
Paratymolus, Haswell, Ann. Mag. Nat. Hist., 1880, Vol. V. p. 302; and Cat. Austr. Crust., p. 142.
Paratymolus, Ortmann, Zool. Jahrb. Syst., &c., VII. 1893-94, p. 34.

I agree with Ortmann in placing this genus among the *Achæus*-like Maiidæ: the position of the external genitalia of an ovigerous female in the Museum collection is conclusive.

Carapace elongate-subpentagonal, not depressed.

Eye-stalks long, slender, salient, non-retractile: no orbits or pre-ocular and post-ocular spines. Antennules longitudinally folded beneath the rostrum.

· Antennæ long, exposed, dorsally, in the greater part of their extent: the basal joint slender, but so short as hardly to reach the front.

19

Rostrum short, emarginate, distinctly delimited from the carapace. Epistome short.

External maxillipeds with the merus narrower than the ischium, and bearing the flagellum at the antero-internal angle.

Legs not elongate : dactyli slender, straight.

Paratymolus hastatus, n. sp.   Plate V. figs. 4, 4*a*.

Carapace somewhat elongate-pentagonal or ovoid, with the rostrum sharply demarcated, and with the regions undefined.

Gastric region with three sharp tubercles disposed in a triangle, base forwards : cardiac region with a single tubercle : branchial regions each surmounted by an oblique crest of 2 or 3, and with a lateral marginal row of 2 or 3, sharp tubercles : hepatic regions each with two sharp lateral teeth, the posterior of which is large. Rostrum short, emarginate, deeply and broadly grooved dorsally.

Eye-stalks long, laterally projecting, slightly moveable forwards but not retractile. Eyes tipped with two or three stiff setæ. No orbits, and nothing in the shape of orbital spines except a slight angular emargination of the base of the rostrum.

Antennæ as long as the post-orbital portion of the carapace, and visible, dorsally, from the base of the second joint of the peduncle : the basal joint, which alone is concealed, although slender is short, hardly reaching the front.

External maxillipeds with the merus broad, but not so broad as the ischium, and giving insertion to the palp at the antero-internal angle.

Trunk-legs with a few coarse stiff setæ : the 2nd pair, which are slightly the longest, are a little less than twice the length of the carapace without the rostrum.

Chelipeds characterized by the carpus, which has its antero-internal angle produced obliquely to form a great spike, the point of which reaches almost to the base of the fingers.

Length of carapace 6 millim.   Breadth of carapace 4·5 millim. Length of 2nd pair of legs 10·5 millim.

An egg-laden female from the Andamans ; in which I am satisfied that the genital orifices are not on the bases of the third pair of legs, but on the sternum.

PHYSACHÆUS, n. gen.

Closely allied to *Achæus*, from which it is distinguished chiefly by the form of the basal joint of the antennary peduncle, which is long and slender, and is fused near its distal end with the tip of the rostrum.

General form that of an *Achæus* with the pterygostomian and branchial regions so inflated as to push forwards the epistomial region to a plane almost at right angles with the antennary region.

Eyes small, slender, rigidly immovable,—in short undergoing degeneration. No orbits or orbital spines.

Rostrum very short, bifid, at tip, the point of each tooth being fused with the distal end of the (otherwise free) sub-cylindrical basal joint of the antennary peduncle. Antennæ of great length.

External maxillipeds with the merus rounded and slightly produced beyond the articulation—at the antero-internal angle—of the palp: the merus much narrower than the ischium. Legs long and slender, with long filamentous dactyli. Chelipeds short.

Physachæus ctenurus, n. sp. Plate III. figs. 2, 2 *a-b*.

Carapace sub-triangular, globosely inflated, with all the regions, except the cardiac, tumid and fairly well delimited, and with a strong post-ocular constriction, beneath which there is an almost vertical descent to the mouth.

The rostrum, which is small, consists of two narrow, slightly divergent, hollow teeth, to either apex of which the distal end of the otherwise perfectly free basal joint of the corresponding antennary peduncle is fused.

Two large erect procurved spines occur in the middle line of the carapace; one on the posterior part of the gastric region, the other behind the cardiac region: on either side of the former, but in a plane anterior to it, there may sometimes be a spinule.

In both sexes the abdomen is bluntly but strongly carinated down the middle line, the carina in the case of the male ending on the 6th tergum in a huge recurved spine: in the female instead of a spine there is a small tubercle, and the posterior edge of the sixth tergum bears a row of four spines.

The eye-stalks are very small, and are rigidly fixed at right angles to the rostrum: the corneæ are almost devoid of pigment. There are no orbits or orbital spines.

The antennæ are distinctly exposed from their base, and are half as long again as the entire carapace, between one-third and two-fifths of their extent being formed by the slender peduncle. The basal joint is slender and almost cylindrical: it is quite free from neighbouring parts, except at the distal end, which is fused with the tip of the rostrum. The flagella are fringed with long hairs.

The antennules are large, and fold longitudinally within the hollow teeth of the rostrum. Except in regard of the fingers, the chelipeds

21

have much the same form as, though slenderer proportions than, those of *Stenorhynchus*, but the merus is much more strongly and elegantly curved : the merus and carpus are moderately inflated, the former joint, like the ischium, having its lower edge more or less granulate : the palm is compressed, with the edges denticulate : the fingers are strongly compressed, and have the cutting edges accurately and completely apposable throughout, being denticulate near the tips only.

In the female the chelipeds have the same general form as in the male, but differ in having the lower edge of the ischium and merus strongly spinate. The legs are slender and filiform, about one-fourth of their length being contributed by the filamentous dactylus : those of the third trunk-segment are the longest, being about four times the length of the carapace, rostrum included, and more than two-and-a-half times the length of the chelipeds.

	Male.	Female.
Length of carapace	7·2 millim.	... 8·5 millim.
Breadth of carapace	6·0 ,,	... 7·0 ,,
Length of legs of 2nd trunk-segment	28·0 ,,	... 28·0 ,,
,, ,, 3rd ,, ,,	32·0 ,,	... 32·0 ,,

Numerous males and egg-laden females from the Andaman Sea, 240 to 375 fathoms.

The eggs are few in number and are singularly large, those from a female of the dimensions given above being over a millimetre in diameter.

Physachæus tonsor, n. sp. Plate III. fig. 3.

The female, which is the only sex represented in the collection, differs from the female of *Physachaeus ctenurus* in the following particulars :—

(1) the gastric region of the carapace, instead of a single large spine, has several smooth tubercles ; and the large spine behind the cardiac region is coarser, and is recurved instead of procurved : the post-ocular constriction is less marked :

(2) the abdominal carina ends in a spine, and the sixth tergum has its after edge perfectly smooth instead of quadrispinate :

(3) the eye-stalks are larger, and are compressed instead of cylindrical :

(4) the chelipeds are relatively stouter, being of much the same proportions as those of the male of *Physachæus ctenurus* : their merus is compressed and has its lower border very strongly and sharply carinated : the hands are much thinner and more compressed ; the palm

22

having its lower edge, and the fingers their outside edges, sharply cristate :

(5) the legs of the second, not of the third, trunk-segment are the longest, and considerably so.

Length of carapace 11 millim. Breadth of carapace 9·5 millim. Length of legs of 2nd trunk-segment 47 millim., of 3rd trunk-segment 40 millim.

Two egg-laden females from the Andaman Sea, 271 fathoms.

The eggs, as in the preceding species, are large and few in number.

The above species represent an *Achæus* modified for life at a considerable depth. The branchial chambers, as is very commonly the case in deep-sea Malacostraca, are greatly inflated : the eyes have degenerated, and the antennæ—no doubt in compensation—have become remarkably lengthened : while the auditory tubercles also, it may be mentioned, are large and prominent.

GRYPACHÆUS, n. gen.

Intermediate between *Achæus* and *Echinoplax*.

Carapace triangular, spiny, separated from the frontal region by a post-ocular "neck." Rostrum spiny : composed of two short divergent spinelets, with a strong median deflexed (interantennulary) spine, not visible from above. Eyes laterally projecting, movable, but not sufficiently retractile to be ever concealed. Small supra-ocular and post-ocular spines are present as part of the general spinature. Antennæ dorsally exposed from the basal joint of the peduncle, which joint is long slender cylindrical and spiny. External maxillipeds with the merus elongate, much narrower than the ischium, and not much broader than the carpopodite. Legs hairy and spiniferous. Abdomen six-jointed in ♀ .

Grypachæus hyalinus (Alcock & Anderson). Plate III. figs. 4, 4*a*.

Achæus hyalinus, Alcock & Anderson, J. A. S. B., Pt. ii. 1894, p. 205.

Carapace sub-triangular, thin, vitreous, spiny especially in its anterior half : the regions well delimited, and the post-ocular portion constricted to form a "neck." The rostrum, as seen from above, ends in two short spines, each of which has a spine at its base ; but from in front or from below it shows a strong vertically deflexed (interantennulary) spine.

The eyes are large ; and the long eye-stalks, which bear two tubercles on their front surface, are movable backwards, and are exposed from

23

their base in all positions. The antennæ are visible, dorsally, from the end of the basal joint of the peduncle, which joint is long, slender, cylindrical and spiny.

The external maxillipeds are large, hairy, and almost pediform, owing to the narrowness of the merus and the coarseness of the palp.

The trunk-legs are hairy and spiny, the hairs on the 2nd and 3rd pairs being remarkably long, stiff, and closely and evenly set. The arm, wrist, and hand of the chelipeds—but especially the arm—are acutely spiny, as are also the edges of the meropodites of the legs,—the spinature of the front edge of the meropodites of the 2nd and 3rd pairs being particularly prominent. The fifth pair of legs are sub-chelate, the propodite having its proximal end strongly dilated to receive the folded-back dactylus : the apposed edge of the dactylus is minutely, that of the propodite sharply and conspicuously, spinate.

Length of carapace 14 millim. Breadth of carapace 9 millim. Greatest span (between extended 2nd pair of trunk-legs) 67 millim.

Loc. Off Trincomalee 28 fms. Females only.

ECHINOPLAX, Miers.

Echinoplax, Miers, " Challenger " Brachyura, p. 31.

Carapace sub-pyriform, longer than broad, and covered with very numerous closely-set spines and spinules : orbital margin spinose : spines of rostrum acute, divergent from their bases, and bearing several accessory spinules. Post-abdomen seven-jointed. Basal antennal joint slender, spinuliferous, and in contact with the front at the distal extremity : flagellum visible from above at the sides of the rostrum. Maxillipeds with the merus narrower than the ischium, and the palp coarse ; merus truncated and not notched at the distal extremity, the antero-lateral angle not produced. Legs spinuliferous. Chelipeds in the female [as in the male] slender and feeble, with the palms not dilated. Ambulatory legs considerably elongated, with the penultimate joint not dilated; the dactyli nearly straight.

Key to the Indian Species of Echinoplax.

Carapace with the regions well defined : rostrum in the adult considerably less than half the length of the carapace :—

1. Carapace and abdominal terga closely covered with pungent acicular spines of equal size...*E. pungens.*
2. Carapace and abdominal terga finely granular, with a few definitely placed spines of conspicuous size ... ... *E. rubida.*

24

Echinoplax pungens, Wood-Mason.

Echinoplax pungens, Wood-Mason, Ann. Mag. Nat. Hist., March, 1891, p. 259.

Carapace pyriform, convex, with the regions well delimited ; densely covered, as are also the sterna, chelipeds, ambulatory legs, and external maxillipeds, with pungent acicular spines. The abdominal terga of the male and young female are also similarly spiny, but in the adult female they become only distantly and coarsely granular.

The rostrum consists of two slender curved divergent spines—less than one-third the length of the carapace proper—the outer and lower surfaces of which are extremely spiny.

The eye-stalks, which have the anterior surface closely spinulate, are retractile, but not to the extent of concealment : there is a strong post-ocular spine—to which, however, the retracted eye does not nearly reach—and numerous smaller spines along the supra-ocular and infra-ocular margins. The antennæ are visible from above, from the middle of the second joint of the peduncle : the peduncle is spiny, with all the joints very slender : the flagellum reaches a little beyond the tip of the rostrum.

The interantennulary spine is large and deeply bifid.

The chelipeds, which are alike in form in both sexes—though relatively longer in the male—are not stouter than the ambulatory legs, and are rather longer than the carapace and rostrum.

The legs of the next pair are more than twice, and those of the third pair rather less than twice the length of the chelipeds, while the fourth and fifth pairs decrease considerably in length : the dactyli of all are densely covered with a brushwork of setæ.

	Male (adult).	Female (adult).
Length of carapace and rostrum ...	70 millim.	79 millim.
Greatest breadth of carapace ...	47 ,,	57 ,,
Length of cheliped	76 ,,	75 ,,
,, ,, 2nd pair ...	158 ,,	191 ,,

Andaman Sea, 130–250 fathoms.

A figure of this fine species has been drawn for " Illustrations of the Zoology of the ' Investigator ' " for 1896.

Echinoplax rubida, n. sp.

Differs from *Echinoplax pungens*, specimens of the same sex, and of approximately the same size being compared, in the following particulars :—

1. The carapace, instead of being everywhere covered with pun-

gent acicular spines of uniform size, is finely granular, with certain definitely placed distant thornlike spines of conspicuous magnitude, namely :—four in triangle on the gastric region, two side by side on the cardiac region, two side by side on the intestinal region, three on each hepatic region, and three on each branchial region : besides these there are some smaller spines on the lateral aspect of the pterygostomian and branchial regions :

2. The rostral spines are less divergent, and have elegantly curved tips :

3. The abdominal terga (of the young female), instead of being everywhere closely covered with pungent spines, are merely finely and distantly granular, with a single large spine on the first tergum, and a pair of smaller spines on the second, in the middle line :

4. The legs are much less spiny, the propodites of the ambulatory legs being fringed with stiff bristles instead of spines :

5. The colour differs, being, in spirit specimens, a warm brown, instead of a pale yellow.

It differs from *Echinoplax moseleyi*, Miers, judging from the figures and description, in the following particulars :—

1. The regions of the carapace are well delimited by sharp cut grooves :

2. The rostral spines are considerably less than half the length of the carapace proper :

3. The armature is altogether different, the large stout spines of the present species standing out on a finely granular carapace, and the abdominal terga being distantly granular.

Total length of carapace 35 millim., breadth of carapace 21 millim., greatest span (2nd pair of trunk-legs) 150 millim.

Loc. Andaman Sea, 90 to 177 fathoms.

PLATYMAIA, Miers.

Platymaia, Miers, ' Challenger' Brachyura, p. 12.

Carapace sub-orbicular. Rostrum short, tridentate owing to the size and projection of the interantennulary septum. No pre-ocular spine ; but a post-ocular spine against which the eye is retractile, but which affords no concealment to the eye. Epistome extremely narrow. Eyes large, with short eye-stalks. Basal antennal joint short, cylindrical, and perfectly free : the flagellum and part of the peduncle visible from above.

External maxillipeds with the meropodite narrow, and bearing the next joint at its summit. Chelipeds in the male long, with a long inflated club-shaped palm: in the female very short and slender. Ambulatory legs long, with remarkably thin compressed joints: some of the legs spiny.

Abdomen in both sexes with all the segments separate.

This genus appears to be very closely related to *Macrocheira*.

Platymaia wyville-thomsoni, Miers.

Platymaia wyville-thomsoni, Miers, 'Challenger' Brachyura, p. 13, pl. ii. fig. 1.
Platymaia wyville-thomsoni, Wood-Mason and Alcock, Ann. Mag. Nat. Hist.,
March, 1891, p. 258, and May, 1894, p. 401.

Carapace transversely sub-circular with the cervical grove well defined: its surface ranging from spinate (in the young) to nearly smooth (in old adults). The rostrum, which is so short as not to break beyond the general outline, consists of three stout spines of equal size, the middle one being the horizontally projecting interantennulary spine.

The hepatic region of the carapace bears (in the adult) a nearly vertically disposed row of three spines, against the upper one of which the eye is retractile.

The eye-stalks are short, and the eyes large and oval. The antennæ are about one-third the length of the carapace, and are plainly visible, in almost the whole of their extent, from above: the joints of the peduncle are short slender and cylindrical, the basal joint being perfectly free.

The external maxillipeds have the meropodite narrow (about half the breadth of the ischiopodite) and giving attachment to the coarse palp at the summit: both meropodite and ischiopodite are spiny.

The chelipeds vary considerably according to sex: in both sexes they are spiny up to the base of the fingers; but whereas in the female and young male they are much slenderer than any of the legs and are not longer than the carapace, in the adult male they are from two to three times the length of the carapace and are much stouter than any of the legs—especially as regards the palm, which is swollen and club-shaped. The 2nd to 5th pairs of legs are long and slender, with the joints thin and compressed, the propodites being blade-like. The 2nd pair, which are from $3\frac{3}{4}$ (female) to $5\frac{1}{2}$ (male) times the length of the carapace, are remarkable for their propodite and dactylus, the front edge of which bears a double comb of enormous spines, the posterior edge also being spinulate: both edges of the merus and carpus also are distantly spinulate. The 3rd and 4th pairs have the front edge of the merus distantly spinulate, and they, as well as the 5th pair, have the front edge of the razor-like merus closely fringed with long stiff hairs.

27

The abdomen in both sexes is seven-jointed, the abdominal terga, like the thoracic sterna, bearing a few spines or tubercles. The epimeral plates corresponding to the third and fourth trunk legs are also spinate.

Andaman Sea, 130–405 fathoms.

A large male of this fine species have been figured for "Illustrations of the Zoology of the ' Investigator ' " for 1896.

Note on some obvious growth-changes in Platymaia wyville-thomsoni.

In very young specimens (carapace less than half an inch in diameter) the whole carapace is closely and sharply spiny.

In larger specimens (carapace about three-quarters of an inch in diameter) the carapace has become closely and finely granular, with the spines persistent only in definite situations, somewhat as in Miers' figure and description (*loc. cit.*)

In larger specimens (carapace two and a half inches in diameter) the carapace has become coarsely and bluntly granular, without any spines, except a few quite anteriorly in the neighbourhood of the hepatic region.

In the largest specimens (carapace three to nearly four inches in diameter) the carapace is in places quite smooth, the only spines present being two external to the eye, and one on the front margin of the hepatic region.

In contrast with the carapace, the spines on the abdominal sterna of the male show no signs of effacement with age.

The colours also vary with age. In young males the carapace is red, with or without white points, and the legs are red and white in alternate bands. In the adult the colour is uniform.

Oncinopus, de Haan.

Oncinopus, de Haan, Fauna Japonica, Crust., p. 87.
Oncinopus, Miers, Journ. Linn. Soc., Zool., Vol. XIV. 1879, p. 645 ; and ' Challenger ' Brachyura, p. 20.

"Carapace semi-membranaceous, elongate, narrow-triangulate and depressed. Rostrum very short, composed of two vertically compressed laminiform lobes: no præ- or post-ocular spines. Post-abdomen in both sexes distinctly seven-jointed. Eyes slender and projecting laterally. Antennæ with the basal joint very short and slender, and not attaining the front, the flagella exposed and visible at the sides of the rostrum. Merus of the exterior maxillipedes elongated, and articulated with the next joint at its summit. Chelipedes in the male rather small, with the palm turgid, and the fingers having between them, when closed, an interspace at the base. Ambulatory legs slender and 'somewhat elongated, with the penultimate joints of the first and second pairs dilated, compressed, and ciliated on the posterior margin; the dactyli in all slightly arcuated and retractile against the penultimate joints."

Oncinopus aranea, de Haan.

Inachus (Oncinopus) aranea, de H., Faun. Japon. Crust., p. 100, pl. xxix. fig. 2.
Oncinopus aranea, Adams and White, Zool. ' Samarang,' Crust., p. 3.
Oncinopus neptunus, Adams and White, Zool. 'Samarang,' Crust., p. 1, pl. ii. fig. 1.
Oncinopus subpellucidus, Stimpson, Proc. Acad. Nat. Sci. Philad., 1857, p. 221.
Oncinopus angulatus, Haswell, Proc. Linn. Soc., N. S. Wales, IV. 1879, p. 483.
Oncinopus subpellucidus, Haswell, Cat. Austr. Crust., p. 5.
Oncinopus aranea, Miers, Zool. 'Alert,' pp. 182 and 190; and ' Challenger ' Brachyura, p. 20.
Oncinopus neptunus, Walker, Journ. Linn. Soc., Zool., Vol. XX. 1890, p. 109.
Oncinopus aranea, Henderson, Trans. Linn. Soc., Zool., (2) V. 1893, p. 341.
Oncinopus aranea, Ortmann, Zool. Jahrb., Syst. etc., VII. 1893, p. 37.
Oncinopus neptunus, Alcock and Anderson, J. A. S. B., Pt. ii. 1894, p. 199.

Carapace elongate-triangular, thin and semi-membranous, and, as well as all the appendages, tomentose. Rostrum short, bilobed.

Eyes small, retractile beneath the edge of the carapace: no orbits or protective spines.

Antennæ extremely short, reaching only just beyond the tip of the rostrum : the basal joint short and free.

Chelipeds in the female and young male slenderer than the next legs and not quite equal in length to the carapace; in the adult male about as stout as the next legs, with an inflated almost globose palm, and a little longer than the carapace.

The 2nd and 3rd pair of legs differ very markedly from the 4th and 5th pair. The 2nd and 3rd pair are long and stout, with a comparatively short carpopodite, with a long broad propodite, and with a comparatively slightly curved dactylus—all these joints being remarkably setaceous. The 4th and 5th pair, on the other hand, are slender and comparatively short, with a long slender carpopodite and with a short propodite which with the strongly recurved dactylus forms a sub-chela—all these joints being merely tomentose. The 5th pair of legs is also remarkable for its sub-dorsal position.

Length of carapace of an adult, 14 to 15 millim.

Specimens in the Museum collection from the Laccadives, Maldives, Ceylon, Andamans and Malay Peninsula, up to 32 fms.

29

CAMPOSCIA, Latreille.

[*Camposcia*, Latreille, Cuvier Regne Animal (2) IV. p. 60.]
Camposcia, Milne-Edwards, Hist. Nat. Crust. I. 282.
Camposcia, de Haan, Fauna Japonica, Crust., p. 87.
Camposcia, Miers, Journ. Linn. Soc., Zool., Vol. XIV. 1879, p. 644.

Carapace pyriform. Rostrum broad, exceedingly short—hardly surpassing the level of attachment of the eyes—emarginate, slightly deflexed.

Eye-stalks long, recurved, retractile towards the sides of the carapace : a post-ocular tooth, not however affording any concealment to the eye. Antennulary fossæ coalescent to form a single chamber. Antennæ moderately long, almost entirely exposed to dorsal view, the free joints of the peduncle flattened.

External maxillipeds with the merus narrower than the ischium, and giving attachment to the next joint at the summit. Chelipeds in both sexes slender—but most so in the female—and short. Some of the ambulatory legs long.

The abdomen in both sexes has all seven joints distinct, and is as broad in the adult male as it is in the adult female— covering almost the whole sternum.

Camposcia retusa, Latr.

[*Camposcia retusa*, Latreille, Cuvier Regne Animal (2) IV. p. 60.]
[*Camposcia retusa*, Guerin, Icon. Regn. Anim. Crust., pl. ix. fig. 1.]
Camposcia retusa, Latr. Milne-Edwards, Hist. Nat. Crust. I. 283, pl. xv. figs. 15 and 16.
Camposcia retusa, Cuvier, Regne Animal, Crust., pl. xxxii. fig. 1.
Camposcia retusa, Adams and White, Zool. ' Samarang,' Crust., p. 6.
Camposcia retusa, Bleeker, Recherches Crust. de l'Ind. Archipel., p. 7.
Camposcia retusa, Stimpson, Proc. Acad. Nat. Sci. Philad., 1857, p. 218.
Camposcia retusa, A. Milne-Edwards, Nouv. Archiv. du Mus., VIII. 1872, p. 255.
Camposcia retusa, Brocchi, Ann. Sci. Nat. (6) II. 1875, Art. 2, p. 89, pl. xviii. fig. 156 (male appendages).
Camposcia retusa, Hilgendorf, Monatsber. Akad. Berl., 1878, p. 784.
Camposcia retusa, Haswell, Proc. Linn. Soc., N. S. Wales, IV. 1879, p. 433; and Cat. Austr. Stalk and Sessile-eyed Crust., p. 4.
Camposcia retusa, E. Nauck, Zeits. Wiss. Zool., xxxiv. 1880, p. 38 (gastric teeth).
Camposcia retusa, Miers, Zool. ' Alert,' pp. 181, 189, 516, and 520.
Camposcia retusa, De Man, Archiv. f. Naturgesch. LIll. 1887, Bd. i. p. 219.
Camposcia retusa, C. W. S. Aurivillius, Kongl. Sv. Vet. Akad. Handl., XXIII. 1888–89, No. 4, p. 35.
Camposcia retusa, A. Ortmann, Zool. Jahrb., Syst., etc., VII. 1893, p. 35.
[*Camposcia retusa*, F. Muller, Verh. Ges. Basel, VIII. p. 473.]

Carapace pyriform, thin, but well calcified. The whole body and

30

most of the appendages thickly setaceous, and densely encrusted with sponges, zoophytes, algæ, etc. Rostrum broad, extremely short, somewhat deflexed, slightly emarginate.

Eye-stalks long, recurved, retractile to the sides of the carapace, and towards a slender acute post-ocular spine. Owing to the imperfection of the rostrum the interantennulary spine is not developed, so that both the antennules fold into a common chamber.

The antennæ, which are completely exposed from the base of the 2nd joint, have the basal joint long and slender, and the free joints of the peduncle flat and densely setaceous.

The hairy external maxillipeds have the antero-internal angle of the ischium produced into a long narrow lobe, parallel to the narrow meropodite.

The chelipeds in both sexes are slender and are about equal in length to the carapace : in the male they are stouter than in the female, and also differ in having the palms inflated : the fingers in both sexes are closely apposable and are toothed throughout.

The other trunk-legs increase in length from the 2nd pair (which are a little longer than the chelipeds) to the 4th pair (which are twice as long as the chelipeds) : the 5th pair, again, being only as long as the 3rd pair.

The abdomen in the adults of both sexes is broad and sub-circular, almost entirely covering the sternum, and consists of seven separate segments.

In the Museum collection are adult males and egg-laden females from the Andamans, Cocos, Ceylon and Samoa—the last being from the collection of the Museum Godeffroy.

Alliance II. INACHOIDA.

INACHOIDES, Edw. & Lucas.

Inachoides, Milne-Edwards and Lucas, in D'Orbigny Voy. Amer. Merid., Crust. pp. 4 & 5.

Inachoides, Miers, Journ. Linn. Soc., Zool., Vol. XIV. p. 646.

Inachoides, A. Milne-Edwards, Miss. Sci. Mex., etc., Crust., etc., I. p. 198.

Carapace pyriform much narrowed in front, inflated behind, the regions well delimited. Rostrum simple. Eyes not, or slightly, retractile towards the sides of the carapace ; never, in any position, concealed. Pre-ocular and post-ocular spines distinct—especially the latter.

Basal antennal joint long and slender : its antero-external angle visible from above, on either side of the rostrum, as an acute spine :

31

the rest of the antennal peduncle, and the flagellum, completely exposed from above.

Epistome broad. External maxillipeds with the merus as broad as the ischium, completely closing the mouth.

Chelipeds in the male rather longer than any of the other legs, and with a long somewhat inflated palm. Ambulatory legs of moderate length, slender, and ending in a styliform dactylus which in some cases is spinulate along the posterior border.

Abdomen of the male composed of seven distinct segments, that of the female of five.

Inachoides dolichorhynchus, Alcock & Anderson. Plate IV. figs. 1, 1a.

Inachoides dolichorhynchus, Alcock and Anderson: Journ. As. Soc., Bengal, Pt. ii. 1894, p. 206.

Carapace elongate-triangular. Rostrum as long as the carapace, simple, spiny, acute. The regions of the carapace are well defined, and are distantly spiny, the following spines being the most conspicuous :— (1) on each side a supra-ocular, a post-ocular (hepatic), and four branchial; (2) in the middle line, a gastric, a cardiac, and an intestinal.

The eyes, though to a certain extent retractile towards the sides of the carapace, are in all positions completely exposed.

The antennæ, which are exposed from the end of the basal joint, are long—more than three-fourths the length of the carapace : their basal joint is long, slender, flattened and fused with the neighbouring parts, and has its antero-external angle produced into an acute spine : the second and third joints are knobbed distally.

The chelipeds are long—one-fourth longer than the carapace and rostrum combined : their palm, which forms about two-fifths of their total extent and is nearly three times the length of the fingers, is broadened and moderately inflated. The 2nd pair of trunk-legs are about equal in length to the chelipeds, but the 4th and 5th pairs are not much more than half that length.

Length of carapace and rostrum 17·5 millim.; greatest breadth 8 millim.; greatest span 54 millim.

Off Madras Coast.

ENCEPHALOIDES, Wood-Mason.

Nearly related to *Inachoides.*

Carapace, owing to the remarkable inflation of the branchial regions, heart-shaped and posteriorly as broad as long (rostrum included) : the branchial regions meeting across the carapace in the middle line. Rostrum simple, shaped like the beak of a bird. Eyes retractile against the sides of the carapace : a small pre-ocular and post-ocular spine, but no definite orbit.

Basal antennal joint slender throughout : the antennæ visible, dorsally, from the base of the second joint.

Merus of the external maxillipeds produced antero-externally to form a foliaceous lobe which covers the greatly produced efferent branchial orifice.

Abdomen in the male seven-jointed : in the female the fourth, fifth and sixth segments, though distinctly recognizable, are firmly fused together.

Chelipeds in both sexes slender. Legs long and slender.

Only eight branchiæ on either side.

Encephaloides armstrongi, Wood-Mason.

Encephaloides armstrongi, Wood-Mason, Ann. Mag. Nat. Hist., March, 1891 p. 259.

Carapace heartshaped : its greatest breadth is equal to its length with the rostrum : its surface in the adult is nodular or pustular, in the young coarsely spiny. The gastric and hepatic regions are well-defined ; but the cardiac and intestinal regions are entirely concealed by the branchial regions, which rise up like a pair of mammæ, and meet, but without any fusion of walls, down the middle line.

The rostrum, which is shaped exactly like the beak of a bird, is about one-fourth the length of the carapace proper, and has a finely serrated edge.

In the male the abdomen is distinctly seven-jointed ; but in the female the fourth, fifth and sixth segments are immovably sutured together.

The eyes which are small, slender, and unpigmented, are retractile against the side of the carapace : there is a very narrow supra-orbital eave ending anteriorly in a minute tooth, and there is a small post-ocular spinule.

On the dorsal aspect the antennæ are plainly visible on either side of the rostrum, from the base of the 2nd joint of the peduncle : the flagella, which are of hairlike tenuity, hardly surpass the tip of the rostrum.

Owing to the prolongation of the efferent branchial canal, the front edge of the buccal frame is V-shaped, and the merus of the external maxillipeds ear-shaped.

33

The trunk-legs recall those of *Egeria*, being all long, slender, cylindrical, and quite devoid of hairs or spines : the chelipeds are short, and are not stouter than the ambulatory legs.

For proportions, see Ann. Mag. Nat. Hist., March, 1891, p. 260.

APOCREMNUS, A. Milne-Edwards.

Apocremnus, A. Milne-Edwards, Miss. Sci. Mex., etc., Crust., etc., I. p. 184.
Apocremnus, Miers, ' Challenger ' Brachyura, p. 17.

Carapace triangular or pyriform, much narrowed in front, inflated behind. Rostrum bifid. Eyes imperfectly retractile : a strong supraocular, but no post-ocular spine [a distant hepatic spine must not be mistaken for a post-ocular spine]. Basal antennal joint narrow, its antero-external angle forming a strong spine visible from above on either side of the rostrum : the free joints of the peduncle and the flagellum exposed to dorsal view. Epistome broad. External maxillipeds with the merus at least as broad as the ischium, quite closing the mouth-frame. Chelipeds not much enlarged : the other legs short and slender, with slender dactyli capable of some flexion on the penultimate joint. Abdomen in the male six jointed—(in the female four (?) jointed).

The genus *Apocremnus* has never yet been reported from Eastern Seas. It was first described from the Florida coast, and was afterwards reported by the ' Challenger ' from Fernando Noronha (an island in the South Atlantic, off the coast of Brazil). There is nothing unprecedented therefore in its occurrence in deepish water in the Indian Ocean.

Apocremnus indicus, n. sp.   Plate IV. figs. 2, 2a.

Carapace pyriform, inflated in the branchial, constricted in the postocular region, and armed with six long knob-headed spines, as follows :—
one, semi-erect, above the root of either eye-stalk ; one in the middle of the cardiac region, flanked on either side by one in the middle of each branchial region ; one in the middle line on the posterior border. There are, in addition, on either side, two sharp spines, one above the other, near the middle of the hepatic region, and far from the eye.

The rostrum is formed of two short, slightly divergent, knob-headed spines. On either side of its base are seen the antennæ and a large spine formed by the antero-external angle of the basal antennal joint.

The constituent segments of the sternum are sharply granular, and are separated from one another by deep grooves.

34

The eye-stalks are of moderate length, salient, and almost immovable. The buccal orifice is large, and the external maxillipeds are ornamented with lines of fine sharp-cut granulation : their merus is as broad as the ischium, and is excavated near the middle for the insertion of the palp. The chelipeds, in the male, are somewhat longer than the carapace and rostrum : their ischium, merus, and carpus are ornamented with lines of fine sharp granulation : the palms are elongate and compressed, with the edges carinate : the fingers, which are less than half the length of the palm, are compressed and curved.

The ambulatory legs, which decrease in length gradually, have their bases and meropodites granular, and the dactyli very slender.

The length of the carapace of the largest specimen—a male—is 9 millim., of an egg-laden female 6 millim.

From off the Andamans at about 100 fathoms, and off Ceylon at 32 to 34 fathoms.

COLLODES, Stimpson.

Collodes, Stimpson, Ann. Lyc. Nat. Hist., New York, Vol. VII. 1862, p. 193.
Collodes, Miers, Journ. Linn. Soc., Zool., Vol. XIV. 1879, p. 645.

Carapace ovate-triangular. Rostrum short, bifid, with the lobes approximate. Eyes of moderate length, retractile against a strong post-ocular process which affords no concealment. Basal antennal joint narrow, a little curved, anteriorly bidentate, one tooth placed behind the other ; mobile part of the antennæ exposed. External maxillipeds with the merus as broad as the ischium, completely covering the mouth. Chelipeds of moderate size. Ambulatory legs short, prehensile, with slender dactyli which in length are equal to their propodites, and are retractile against the latter. Abdomen of the female consisting of five segments.

Collodes malabaricus, n. sp. Plate V. fig. 3.

Carapace ovate-triangular, with the gastric and cardiac regions distinct and elevated. Rostrum short, emarginate. Pre-ocular spine large and coarse, post-ocular spine very prominent. A tubercle on the cardiac region, and a large epibranchial spine on either side of it.

Basal antennal joint narrow throughout, and bearing two spines anteriorly—one at the antero-external angle, visible from above, and comparable in size to one of the rostral teeth—and one behind this, immediately in front of the base of the eye-stalk. Eyes slender and

35

retractile towards the post-ocular tooth, which, however, affords no concealment.

Chelipeds (in the female) hardly stouter than the ambulatory legs, which are short, with prehensile dactyli.

Two ovigerous females, the larger of which is 4 millim. long, from off the Malabar Coast, 26 to 31 fathoms.

The genus *Collodes* has hitherto been known only as a tropical American genus. It has been found on both sides of Central America so that its occurrence in Indian waters is not without precedent.

Sub-family II. ACANTHONYCHINÆ.

Eyes without true orbits: eye-stalks little movable, either short and more or less concealed beneath a forwardly-directed supra-ocular spine, or obsolescent and almost or completely sunk either in the sides of a huge beak-like rostrum, or between low pre-ocular and post-ocular excrescences (*Sphenocarcinus*) : a distinct post-ocular spine, which is not cupped, may be present (*Pugettia*). Basal antennal joint truncate-triangular.

External maxillipeds with the merus as broad as the ischium, and with the (small) palp arising from the antero-internal angle of the merus.

Dactyli of the ambulatory legs prehensile or sub-chelate, in the former case the last three pairs of legs are often disproportionately short compared with the second pair. Rostrum either simple or two-spined.

Key to the Indian genera.

I. Rostrum of huge size ; simple, or bifid at tip; not flanked on either side by salient supra-ocular spines.	1. Eye-stalks almost obsolete, completely sunk, and almost or quite immovable: carapace smooth or tuberculate: no post-ocular process.	i. Carapace and rostrum sub-cylindrical, the latter bifid at tip..........	XENOCARCINUS.
		ii. Carapace depressed, elongate-triangular: rostrum laterally compressed, not bifid at tip...	SIMOCARCINUS.
	2. Eye-stalks short, sunken but movable between low smooth pre-ocular and post-ocular excrescences : carapace with huge symmetrical pedicled tablets........		SPHENOCARCINUS.

II. Rostrum flanked on either side by salient supra-ocular spines; either long and simple, or consisting of two spines of moderate length : no post-ocular process.

1. Carapace elongate-triangular, rostrum elongate, simple : ambulatory legs not subchelate.

   i. Rostrum laterally compressed, supra-ocular spines small : eye-stalks so short and deeply sunken as to hardly reach to the sides of the carapace ; carapace of the female with large foliaceous lateral lobes........ HUENIA.

   ii. Rostrum horizontally compressed, supra-ocular spines large : eye-stalks short, but reaching beyond the sides of the carapace : carapace of the female without foliaceous lobes................ MENÆTHIUS.

2. Carapace broad, sub-quadrangular : rostrum short and deeply bifid, ambulatory legs subchelate... ACANTHONYX.

XENOCARCINUS, White.

Xenocarcinus, White, Jukes' Voyage H. M. S. ' Fly,' Vol. II. p. 335.
Huenioides, A. Milne-Edwards, Ann. Soc. Entomol. France (4) V. 1865, p. 144.
Xenocarcinus, Miers, Journ. Linn. Soc., Zool., Vol. XIV. 1879, p. 648, pl. xii. fig. 5.

Carapace ovate-subcylindrical, tapering to a long thick subcylindrical rostrum, or beak, the tip of which is emarginate or bifid.

Eyes short, completely sunken in the sides of the rostrum, almost immovable : no præ-ocular or post-ocular spines.

Antennæ with the basal joint triangular, and with the short mobile portion hidden beneath the rostrum.

External maxillipeds with the merus as broad as the ischium and giving attachment to the palp at its antero-internal angle.

Chelipeds not much shorter or stouter than the 2nd and 3rd pairs of legs : 4th and 5th pairs of legs short : all with the dactyli short, stout, curved, and sharply toothed along the posterior surface.

Abdomen of the female four-jointed, the 3rd — 6th segments being fused together.

37

Xenocarcinus tuberculatus, White.

Xenocarcinus tuberculatus, White, P. Z. S., 1847, p. 119, and Ann. Mag. Nat. Hist. (2) I., 1848, p. 221, and in Jukes' Voyage H. M. S. 'Fly,' Vol. II. p. 336.

Xenocarcinus tuberculatus, Hess, Archiv. f. Naturges. XXXI. i. 1865, pp. 131 and 171.

Xenocarcinus tuberculatus, A. Milne-Edwards, Nouv. Archiv. du Mus. VIII. 1872, p. 253, pl. xii. fig. 1.

Xenocarcinus tuberculatus, Miers, Zool. 'Erebus' and 'Terror,' Crust., p. 1, pl. ii. fig. 1, 1e.

Xenocarcinus tuberculatus, Haswell, P. L. S., N. S. Wales, Vol. IV. 1879, p. 436, and Cat. Austr. Crust., p. 8.

Xenocarcinus tuberculatus, Ortmann, Zool. Jahrb. Syst., etc., VII. 1893, p. 40.

Carapace elongate ovate-subcylindrical with the regions ill defined and the surface more or less tuberculated. [Typically the tubercles fall into distinct transverse rows]. The rostrum has the form of a long coarse cylindrical beak, the apex of which is bifid, and the surface densely covered with velvety hairs.

The eyes are completely and almost immovably sunk in the sides of the rostrum.

The antennary flagella are much shorter than, and are completely hidden by, the rostrum.

The chelipeds and ambulatory legs are short and nodular, the latter having curved strongly-toothed prehensile dactyli. The chelipeds are hardly stouter, and are not much shorter, than the 2nd pair of legs, which again are much longer than the 3rd to 5th pair. The colours described by White are "two or three waved longitudinal red lines on the posterior half of the carapace, the inner line continued before the eyes." By A. Milne-Edwards the colours of the carapace and legs are said to be reddish stained with yellow.

In a good spirit specimen the abdomen carapace and beak are dull reddish brown, with a broad yellow stripe extending from the base of the beak to the tip of the abdomen, and on either side of the carapace a narrow sinuous yellow line; and the trunk-legs are yellow, more or less banded and striped with dull brown.

In the Museum collection are two females, one from Ceylon (34 fathoms), the other from the Andamans. The one from Ceylon, which is an egg-laden adult 15 millim. long, resembles as to its carapace and rostrum, but not as to its legs, the figure in the Zoology of the 'Erebus' and 'Terror;' and as to its legs, but not as to its carapace and rostrum, the figure in Archiv. du Mus. tom. VIII. 1872. The other, from the Andamans, which is not adult, exactly resembles, as to its carapace, but not as to its legs, the last cited figure.

SPHENOCARCINUS, A. Milne-Edwards.

Sphenocarcinus, A. Milne-Edwards, Miss. Sci. Mex., Crust., I., p 135.
Sphenocarcinus, Miers, Journ. Linn. Soc., Zool., Vol. XIV. 1879, p. 663; and
' Challenger ' Brachyura, p. 84.

Carapace elongate sub-pentagonal, broad behind, tapering in front to a long rostrum formed of two spines (fused together to near the tip). The surface of the carapace is symmetrically and deeply honey-combed by broad deep channels which leave symmetrical tubercles with over-hanging edges between them.

There are no true pre-ocular and post-ocular spines, but the eye is deeply sunk between two low smooth excrescences which are pre-ocular and post-ocular in position.

The basal antennal joint is truncate-triangular, and the antennary flagella are completely hidden beneath the rostrum. The epistome is long and narrow. The external maxillipeds have the merus as broad as the ischium, somewhat dilated at the antero-external angle, and somewhat excavated at the antero-internal angle for the insertion of the small palp. The chelipeds are not much stouter, and not much shorter than the next pair of legs, which are the longest: the dactyli of the legs, though stout recurved and prehensile, are not toothed along the posterior edge. Abdomen, in both sexes, seven-jointed.

Oxypleurodon Miers (' Challenger ' Brachyura, p. 38) differs from *Sphenocarcinus* only in the form of the rostrum, the spines of which are divergent instead of convergent and more or less fused. I much suspect the generic value of this character. If, however, the two forms be identical, then *Sphenocarcinus* would have to be removed to the next sub-family, in which case the sub-family Acanthonychinæ would be perfect-ly homogeneous.

Sphenocarcinus cuneus (Wood-Mason).

Oxypleurodon cuneus, Wood-Mason, Ann. Mag. Nat. Hist., (6) VII. 1891, p. 261.

Carapace elongate sub-pentagonal, narrowing to a long tapering cylindrical rostrum, which, in the male, is longer than the carapace and only emarginate at the extreme tip, but, in the female, is shorter than the carapace and distinctly bifid at the end.

The carapace is symmetrically honey-combed by deep channels, which leave between them great symmetrically undermined islets, as follows :—one, very elongate-oval, on the gastric region ; one, triangu-lar, on the cardiac region ; one, somewhat semilunar with one horn

39

much produced laterally, on each branchial region ; and one, Cupid's bow-shaped, along the posterior border. Besides these there are some smaller islet-like excrescences, namely, on each side, a supra-ocular, post-ocular, hepatic, and branchial.

Between the supra and post-ocular excrescences, are set the small squat little-movable eyes.

Of the trunk-legs, the 2nd pair (*i.e.*, first ambulatory legs) are the longest, being very slightly longer than the chelipeds, and considerably shorter than the carapace measured with the rostrum, but much longer than any of the last 3 pairs of legs.

In the female all the long joints, except the dactyli, and in the male all except the dactyli and propodites, are strongly carinated dorsally.

The chelipeds are hardly stouter than the next pair of legs, except as regards the palm in the male, which is broadened and somewhat inflated. In neither sex are the short white polished fingers apposable throughout.

	Male.	Female.
Length of carapace and rostrum	... 19· millim.	... 18·5 millim.
Greatest breadth of carapace	... 12· ,,	... 13· ,,
Length of rostrum alone	... 10·5 ,,	... 8·7 ,,
,, of 2nd pair of trunk-legs	... 15·5 ,,	... 15· ,,

Loc. Andaman Sea, 161 to 250 fathoms.

This extremely elegant species has been figured for next year's issue of " Illustrations of the Zoology of the ' Investigator.' "

HUENIA, de Haan.

Huenia, de Haan, Faun. Japon. Crust., p. 83
Huenia, Miers, Journ. Linn. Soc., Zool., Vol. XIV. 1879, p. 648 ; and ' Challenger ' Brachyura, p. 34.

Carapace depressed, elongate-triangular in the male,[*] with the lateral epibranchial angles produced ; sub-quadrangular in the female, with two large foliaceous lobes (epibranchial and hepatic) on either side : a small pre-ocular, but no post-ocular spine. Rostrum simple, acute, vertically deep, laterally compressed. Abdomen in the male seven-jointed ; in the female five-jointed ; with the fourth to the sixth joints coalescent.

Eyes very small and almost immobile.

[*] A small hepatic lobe is sometimes present in the male also, on either side.

Basal antennal joint somewhat enlarged, and coalescent at its distal extremity with the front; beneath which the flagella are inserted out of sight in a dorsal view.

The external maxillipeds are small, the merus distally truncated, and bearing the palp at its antero-internal angle. Chelipeds in the male moderately developed, with the palms compressed and cristate above, the fingers somewhat excavated at the tips, and not apposable throughout their extent. Ambulatory legs short—the longest pair not much longer than the chelipeds, dactyli short, stout, strongly recurved, and more or less toothed along the posterior margin.

Huenia proteus, de Haan.

Maja (Huenia) proteus, de Haan, Faun. Japon. Crust., p. 95, pl. xxiii. figs. 4–6.

Huenia proteus, Adams and White, 'Samarang' Crustacea, p. 21, pl. iv. figs. 4–7, and p. 22, pl. iv. fig. 5.

Huenia proteus, Haswell, Proc. L. S., N. S. Wales, Vol. IV. 1879, p. 437; and Cat. Austr. Crust, p. 9.

Huenia proteus, Miers, Zool. 'Alert,' pp. 182 and 191, and 'Challenger' Brachyura, p. 35.

Huenia proteus, C. W. S. Aurivillius, Kongl. Svensk. Vet. Akad. Handl. XXIII. 1888-89, No. 4, p. 40, pl. iii. fig. 3.

Huenia proteus, R. I. Pocock, Ann. Mag. Nat. Hist. (6) V. 1890, p. 79.

Huenia proteus, Henderson, Trans Linn. Soc., Zool. (2) V. 1893, p. 341.

Huenia proteus, Ortmann, Zool. Jahrb., Syst., etc., VII. 1893, p. 40.

Carapace flat, depressed, with two low elevations in the middle line, otherwise smooth : in the male the carapace is elongate triangular, with the lateral epibranchial angles produced to form small lobes, and sometimes with the hepatic regions expanded in the same way : in the female the carapace is quadrilobate, owing to the foliaceous extension of the hepatic and epibranchial angles. Rostrum long, simple, acute, deep, and laterally compressed. Supra-ocular spines small. Eyes small, deeply sunk beneath the pre-ocular spine, almost immovable.

In the male the chelipeds are somewhat shorter, and the next pair of legs (which are the longest) are somewhat longer than the carapace and rostrum combined: in the female the chelipeds are considerably shorter than, and the next pair of legs are about the same length as, the carapace and rostrum. In the female and young male the fingers, which are closely toothed, meet throughout the greater part of their extent : in the male they meet only at the tips.

The last three pairs of legs are very short. All the long joints, except the dactyli, of all the trunk-legs are more or less carinate dorsally (anteriorly), the carination often being more or less discontinuous in the case of the chelipeds: the dactyli of the ambulatory legs are stout, strongly recurved, and more or less toothed along the posterior margin.

In the Museum collection there are several females, but only two males, from various parts of the Andamans, up to 20 fathoms.

SIMOCARCINUS, Miers.

Simocarcinus, Miers, Journ. Linn. Soc., Zool., Vol. XIV. 1879, p. 649.

As *Huenia*, but without the supra-ocular spine; with the chelipeds much stouter, especially as to the palm, which is much inflated; and with the ambulatory legs more cylindrical.

Simocarcinus pyramidatus (Heller).

Huenia pyramidata, Heller, Crust. Roth. Meer., in SB. Akad. Wien XLIII. 1861 p. 307, pl. i. fig. 9.

Description of the Male.

Carapace elongate-triangular, narrowing to a huge, deep, laterally compressed rostrum of greater length than the carapace: the hepatic regions are marked by a faint bulge, and the lateral epibranchial angles are very sharp cut, while the limits of the posterior border are bounded on either side by a small lobule. Except for a somewhat elongate eminence on the gastric region and a tubercle on the posterior cardiac region, the carapace is perfectly smooth.

The eyes are deeply sunk, and nearly immobile, and the cornea is somewhat deficient in pigment.

The chelipeds, which are markedly stouter than the other legs, are a little shorter than the carapace and rostrum; and the next pair of legs, which are a good deal more than twice the length of the 3rd pair and than thrice the length of the 5th pair, are equal in length to the carapace and rostrum. The palms are broadly inflated; and the fingers, which are strongly arched, meet only at the tips.

The ambulatory legs are cylindrical, and their dactyli are stout, strongly recurved, and toothed along the posterior margin.

Our single perfect specimen—a male from the Nicobars—measures 30 millim. in length of carapace and rostrum.

Simocarcinus simplex (Dana).

Huenia simplex and *brevirostrata*, Dana, U. S. Expl. Exp. Crust. I. pp. 133 and 134, pl. vi. figs. 3a–c, 4a–c.

Simocarcinus simplex, Miers, Jour. Linn. Soc., Zool., Vol. XIV. 1879, p. 649; and 'Challenger' Brachyura, p. 35 (*ubi synon.*).

[*Simocarcinus simplex*, Cano, Boll. Soc. Nat. Napol. III. 1889, p. 173.]

Simocarcinus simplex, J. R. Henderson, Tr. Linn. Soc. Zool. (2) V. 1893, p. 342.

This species is distinguished from *Simocarcinus pyramidatus* (Hell.) (1) by the much shorter rostrum of the male; (2) by the presence of

42

three tubercles, disposed in a triangle, on the gastric region; (3) by
the larger and more prominent eyes; (4) by the absence of the lobule
on either side of the posterior border of the carapace; (5) by the much
more massive chelipeds of the male.

This species is included in the Indian Fauna on the authority of
Prof. J. R. Henderson. There are no specimens in the Indian Museum.

MENÆTHIUS, Edw.

Menæthius, Milne-Edwards, Hist. Nat. Crust. I. 338.
Menæthius, Miers, Journ. Linn. Soc., Zool., Vol. XIV. 1879, p. 649; and 'Challenger' Brachyura, p. 36.

Carapace subpyriform, moderately convex, and tuberculated on
the dorsal surface, with a large triangulate præ-ocular spine, but no
post-ocular spine. Rostrum simple, slender, acute, or emarginate at
apex. Post-abdomen in the male seven-jointed, in the female usually
five-jointed, the penultimate joint formed by the coalescence of three
segments. Eyes small, mobile, but not perfectly retractile. Basal
antennal joint slightly wider at the base than at the distal extremity,
which is unarmed; flagellum exposed and visible from above at the
side of the rostrum. Merus of the exterior maxillipedes truncated at
the distal extremity and with a prominent antero-external angle, and
slightly notched at the antero-internal angle where it is articulated with
the next joint. Chelipedes (in the male) well developed, with the palm
slightly compressed; fingers acute, and having between them, when
closed, an interspace at the base. Ambulatory legs of moderate length;
the joints subcylindrical, not dilated or compressed; dactyli slightly
curved and partially retractile. (Miers).

Menæthius monoceros, (Latr.) Edw.

[*Pisa monoceros*, Latr., Encycl. X. 139.]
Inachus arabicus, Rüppell, Krab. Roth. Meer., p. 24, pl. v. fig. 4.
Menæthius monoceros, Milne-Edwards, Hist. Nat. Crust., Vol. I. p. 339.
Menæthius subserratus, porcellus, and *tuberculatus*, Adams and White, 'Samarang'
Crustacea, pp. 18 and 19, pl. iv. figs. 1 and 2.
Menæthius angustus, depressus, subserratus, tuberculatus, areolatus and *inornatus,*
Dana, U. S. Expl. Exped., Crust. I. pp. 121–125, pl. iv. figs 5a–7g, and pl. v.
figs. 1a–3d.
Menæthius subserratus, dentatus and *depressus,* Stimpson, Proc. Ac. Nat. Sci.
Philad., 1857, p. 219.
Menæthius monoceros, Heller, Crust. Roth. Meer., SB. AK. Wien, XLIII. 1861,
p. 306.
Menæthius monoceros, A. Milne-Edwards in Maillard's L'ile Réunion, Annexe F,
p. 6; and *rugosus* p. 7, pl. xvii. fig. 2.
MENÆTHIUS MONOCEROS, A. MILNE-EDWARDS, NOUVELLES ARCHIVES DU MUSEUM
IV. 1868, p. 70, and VIII. 1872, PP. 252 and 253 (UBI. SYNON.)

43

Menæthius monoceros, Miers, Phil. Trans. Vol. 168, 1879, p. 485, and Zoology 'Alert,' pp. 182, 190, 517 and 521, and 'Challenger' Brachyura, p. 37.
Menæthius monoceros, Haswell, P. L. S., N. S. Wales, Vol. IV. 1879, p. 437, and Cat. Austr. Crust., p. 9.
Menæthius monoceros, de Man, Notes Leyden Mus. II. 1880, p. 171, and Archiv. f. Naturges. LIII. 1887, i. 219.
Menæthius monoceros, Richters in Möbius Meeresf. Mauritius, p. 145.
[*Menæthius monoceros*, Cano. Boll. Soc. Nat. Napol. III. 1889, p. 175.]
Menæthius monoceros, Henderson, Trans. Linn. Soc. Zool. (2) V. 1893, p. 342.
Menæthius monoceros, Ortmann, Zool. Jahrb. Syst., etc., VII. 1893, p. 41.

Carapace elongate-triangular, most markedly so in the male, the lateral epibranchial angles sharp-cut, and the surface very variably tuberculated.

The rostrum, which is flanked on either side by the forwardly-directed supra-ocular spine, is styliform, acute, and horizontally compressed, its length being about half that of the carapace in the male, but a good deal less in the female.

The small eyes are imperfectly retractile, and project freely from beneath the supra-ocular spine.

The chelipeds in the male are as long as, or a little longer than, the 2nd pair of legs, or about equal in length to the carapace and rostrum : they are very much stouter than any of the other legs, and have a somewhat inflated palm, and fingers which meet only at the tips.

The chelipeds in the female are not stouter than the other legs, and are considerably shorter than the next pair of legs, which, again, are a good deal shorter than the carapace and rostrum : the fingers meet through the greater part of their extent.

The 3rd–5th pair of legs are very much shorter than the 2nd pair : in all the dactyli are strongly recurved and are toothed along the posterior margin.

Very numerous specimens from the Andamans and Nicobars.

ACANTHONYX, Latr.

[*Acanthonyx*, Latreille, Regne Animal, (2) IV. 58.]
Acanthonyx, Milne-Edwards, Hist. Nat. Crust. I. 342.
Acanthonyx, A. Milne-Edwards, Miss. Sci. Mex., Crust. I. 142.
Acanthonyx, Miers, Journ. Linn. Soc., Zool., Vol. XIV. 1879, p. 650; and 'Challenger' Brachyura, p. 42.

Carapace sub-oblong, rounded behind, and with the dorsal surface usually depressed, not markedly constricted behind the prominent antero-lateral angles, the lateral branchial spines small and not prominent. Præ-ocular spine prominent, acute. Spines of the rostrum united at the base, acute and but little divergent. Post-abdomen in the male six-jointed. Eyes small, mobile, but not completely retractile. Basal an-

44

tennal joint narrowing slightly from the base to the distal extremity, which is unarmed ; flagellum exposed and visible from above at the side of the rostrum. Merus of the exterior maxillipeds truncated at the distal extremity and but slightly notched at the antero-internal angle, where it is articulated with the next joint. Chelipeds (in the adult male) well developed ; palm compressed, but slightly turgid in the middle, and often slightly carinated above ; fingers acute, and having between them, when closed, an interspace at the base. Ambulatory legs short, with the penultimate joints more or less dilated and compressed and armed with a tooth or lobe on its inferior margin, against which the small acute dactylus closes. (Miers).

Acanthonyx macleayi, Krauss.

Acanthonyx macleayi, Krauss, Sudafrikan. Crust., p. 47, pl. iii. fig. 6.
Acanthonyx macleayi, Miers, 'Challenger' Brachyura, p. 43.

Carapace sub-quadrangular, with the hepatic and lateral branchial spines well developed : these spines, as well as the spines of the rostrum and the carapace immediately behind the rostrum, are tufted with setæ ; and on the gastric region in a line with the hepatic spines are two elevated tufts of setæ. Except for the spines and elevations abovementioned, and for a slight median elevation in its posterior half, the carapace, both as to its margins and as to its surface, is perfectly smooth and unarmed.

The supra-ocular spines are parallel with, and in the female almost comparable in size with the rostral spines.

The chelipeds in the male, but not in the female, are much stouter than any of the other legs : in the male they are nearly as long as the carapace, and have the carpus and palms much inflated, and the fingers in contact only at their tips : in the female they are only about two-thirds the length of the carapace, and have the joints slender, and the fingers closely apposable throughout.

The other legs, which are subchelate, are not disproportionately short compared with the chelipeds : the last pair is sub-dorsal in position.

In the Museum collection are specimens from Karáchi.

Acanthonyx consobrinus, A. Milne-Edwards.

Acanthonyx consobrinus, A. Milne-Edwards, in Maillard's l'Ile de la Réunion, Annexe F. p. 7, pl. xvii. figs. 3, 3b.
Acanthonyx consobrinus, Heller, 'Novara' Crustacea, p. 5.

" Carapace broadened, and a little swollen, surface non-granular. Gastric region with three ill-defined tubercles. Cardiac region either smooth or with sometimes a trace of a rudimentary tubercle. Latero-

45

anterior border cut into four or five teeth, of which the first, or external orbital angle, is small and pointed, the second larger *et à extrémité mousse*, and the others successively smaller. The rostrum consists of two short stout spines, and the supra-ocular border forms a spine. Chelipeds short: fingers evenly toothed. Ambulatory legs ending in a recurved claw. The abdomen of the male consists of 5 segments, the 2nd, 3rd and 4th being fused together.

There are no specimens of this species in the Museum Collection, which is included in this Fauna on the authority of Dr. Heller who mentions it in the ' Novara ' Collection, from Madras.

The genus or sub-genus *Scyramathia* has, I think, very close affinities with the genus *Pugettia*, and is certainly, I think, a close link between this sub-family and the following.

Sub-family iii. PISINÆ.

Eyes with commencing orbits, of which one of the most characteristic parts is a large, blunt, usually isolated and cupped post-ocular tooth or lobe, into which the eye is retractile, but never to such an extent as to completely conceal the cornea from dorsal view: there is also almost always a prominent supra-ocular eave, the anterior angle of which is sometimes produced forwards as a spine. Eye-stalks short. Basal antennal joint broad, at any rate at the base; its anterior angle generally produced to form a tooth or spine. Merus of the external maxillipeds, owing to the expansion of its antero-external angle, broader than the ischium, and carrying the palp at its antero-internal angle. Rostrum two-spined (in *Doclea* obscurely so). Legs often very long.

Key to the Indian Genera.

Alliance 1. PISOIDA. Supra-ocular eave not in close contact with the post-ocular spine or process, and generally produced, but not very conspicuously, at the antero-external angle in the plane of the rostrum.

I. Spines of the rostrum s e p a r a t e from the base, usually long and divergent.

1. Post-ocular tooth either not cupped, or if cupped then the carapace is armed with long acute spines of uniformly large size and regular arrangement ...... SCYRAMATHIA.

2. Post-ocular tooth deeply cupped; spines of the carapace, if present, never of uniform size and arrangement.

i. Spines of the rostrum bearing a secondary spinule, either at tip or somewhere in their distal half .... NAXIA.

ii. Spines of the rostrum without a secondary spinule HYASTENUS.

46

1. Carapace sub-circular or globular: rostrum emarginate: ambulatory legs of moderate length, stout: the entire body, and the appendages in great part, densely tomentose .... DOCLEA.

II. Spines of the rostrum coalescent in their basal half.

2. Carapace broadly triangular: tip of the rostrum deeply cleft: ambulatory legs extremely long and slender.

i. Post-ocular lobe completely isolated both from the supra-ocular eave and from the basal antennal joint: 2nd pair of trunk-legs never approaching six times the length of the carapace... CHORILIBINIA.

ii. Space between the post-ocular lobe and the supra-ocular eave, as well as that between the post-ocular lobe and the basal antennal joint occupied by a spine: 2nd pair of trunk-legs six or more times the length of the carapace............... EGERIA.

Alliance 2. LISSOIDA. Supra-ocular eave in the closest contact with the post-ocular process, and with its antero-external angle almost always (always in Indian genera) very strongly produced forwards in the plane of the rostrum.

i. Surface of carapace tubercular: chelipeds of the male stouter than those of the female: abdomen of the female seven-jointed..................................... TYLOCARCINUS.

ii. Surface of carapace spiny: chelipeds of the male not stouter than those of the female: abdomen of the female five-jointed........... . ... HOPLOPHRYS.

Alliance I. PISOIDA.

SCYRAMATHIA, A. Milne-Edwards.

Scyramathia, A. Milne-Edwards, Compt. Rend. XCI. 1881, p. 356.
Scyramathia, Sars, Norwegian North-Atlantic Expedn., Crustacea Iᴬ. p. 5.
Scyramathia, S. I. Smith, 'Albatross' Crustacea (1884), 1886, p. 21.
Anamathia (part) Miers, 'Challenger' Brachyura, p. 25.

Carapace pyriform or elongate-triangular, armed either with tubercles, or with long spines much like those of *Anamathia* in their uniform size and definite arrangement: the hepatic and lateral epi-

47

branchial spines are always prominent and very conspicuous. The rostrum consists of two spines, which are usually long and slender. The eyes are small, and are retractile against a sharp post-ocular process which commonly is but little cupped : there is also a supra-ocular eave which terminates either in a forwardly directed tooth or in an upturned spine. Basal antennal joint not very broad, sharply truncated : the mobile portion of the antennæ freely exposed on either side of the rostrum.

Merus of the external maxillipeds as broad as the ischium, slightly expanded at the antero-external angle, and bearing the palp at the antero-internal angle.

Chelipeds in the adult male (but not in the female and young male) enlarged, with the palms broadened and compressed.

First pair of ambulatory legs markedly the longest.

The abdomen in both sexes consists of seven distinct segments.

There is certainly a close superficial resemblance between this genus and *Anamathia ;* but I quite agree with Prof. Sars that the two forms are not very closely united. Prof. Sars thinks that *Scyramathia* is nearest to *Hyastenus,* an opinion with which I concur, although I also think that there are quite as close relations to *Pugettia.*

Scyramathia pulchra, Miers.

Anamathia pulchra, Miers, ' Challenger ' Brachyura, p. 26, pl. iv. fig. 1 (adult male).

Anamathia livermorii, Wood-Mason, Ann. Mag. Nat. Hist. March 1891, p. 260 (young male and adult female).

Body and limbs everywhere closely covered with short hairs, which on the carapace are peg-shaped ; and with numerous long scattered setæ. The carapace, which is subpyriform, is armed with twenty long sharp spines disposed in five longitudinal series. Of these spines five are on the gastric region, one is on the cardiac, and one on the intestinal region, one stands above either eye, one on each hepatic, and four on each branchial region : in addition there is a distinctly cupped post-ocular lobe.

The rostrum consists of two slender divergent spines, the length of which is more than half that of the carapace.

The eyes are small, and the cornea, though retractile against the post-ocular lobe, can never be concealed.

The basal antennal joint is broad, and has its antero-external angle somewhat produced : the mobile portion of the antenna is completely exposed to dorsal view.

48

The external maxillipeds have the ischium and merus somewhat concave.

The chelipeds vary according to sex. In the adult male they are longer than the carapace and rostrum, and are far stouter than any of the other legs : the carpus is enlarged and sculptured, the palm is broadened, as well as somewhat carinate along both edges and strongly produced at the postero-inferior angle, and the fingers are opposable in their distal half only : in the female and young male they are shorter than the carapace with the rostrum, and are hardly stouter than the other legs ; all the joints are subcylindrical, and the fingers are apposable in the greater part of their extent.

In both sexes, the merus of all the legs, including the chelipeds, has a spine or tooth at the far end of its upper margin. The 2nd pair of trunk-legs, which are the longest, are, in the male, nearly twice the length of the carapace and rostrum, but in the female are considerably shorter.

Loc. Andaman Sea, 130 to 561 fathoms.

Scyramathia rivers-andersoni, n. sp.

Carapace closely covered with peg-shaped hairs with long setæ interspersed : legs with few setæ. The carapace, which is pyriform and somewhat inflated, has, besides a supra-ocular tooth and a sharp post-ocular process, and besides a salient hepatic spine, and a still more salient lateral epibranchial spine (about two-fifths the greatest breadth of the carapace in length) six sharply conical tubercles evenly and equidistantly arranged in a circle round a central caradiac tubercle : of these the most posterior overhangs the middle of the posterior border, while the most anterior, which is situated far back on the gastric region, is flanked on either side by a very faint eminence.

The rostrum consists of two slender divergent horns, the length of which in the male is about three-quarters, in the female about two-thirds, that of the rest of the carapace.

The eyes are small, and though freely movable forwards are not retractile backwards further than to impinge against the summit of the post-ocular process of the carapace. The basal antennal joint, which is of no great width, is sharply truncated : the mobile portion of the antenna is freely exposed on either side of the rostrum.

The chelipeds in the fully adult male (but not in the young male) are much stouter than the other legs, and are as long as the carapace and rostrum ; their merus is prismatic with knife-like edges, the upper edge ending in a spine ; their carpus is bicarinate, the outer carina being very prominent ; the hands, which form nearly half their total

49

length, have the palm carinate along the upper edge, and the fingers slightly separated when closed.

In the female the chelipeds are not stouter than the other legs, are not much longer than the carapace proper, and have the fingers closely apposable throughout.

Of the ambulatory legs the first are much the longest, being nearly half again as long as the carapace and rostrum; while the last two pairs are very short and have their dactyli reduced in length, increased in strength, and strongly recurved.

		Male.		Female.	
Length of carapace and rostrum...		21	millim.	16·5	millim.
,,	rostrum	9	,,	7	,,
,,	chelipeds	21	,,	11	,,
,,	2nd pair of trunk-legs...	31	,,	20	,,
,,	5th	15	,,	11	,,

Loc. Off Malabar coast, 406 fms.

Scyramathia beauchampi (Alcock and Anderson).

Anamathia beauchampi, Alcock and Anderson, J. A. S. B., 1894, Pt. ii. p. 185.

Body and legs downy, and with numerous large coarse curly clavate hairs, which are very regularly arranged on the legs, where also they are coarsest and closest. Carapace sub-triangular, with the following armature :—

On either hepatic region a great up-curved earlike spine (without any bullous base). On either branchial region, posteriorly, a strong up-turned spine; and anteriorly, near the middle line, a smaller coarse tooth. On the gastric region four sharpish tubercles. On the narrow sunken cardiac region a coarse sharp tooth. On the posterior border, in the middle line, a coarse granule.

The rostrum consists of two more (♀) or less (♂) divergent spines, the length of which is about one-third that of the rest of the carapace.

The eyes are small, and are almost devoid of pigment: they are to some extent hidden beneath a pre-ocular tooth of moderate dimensions, and are retractile against a larger laterally-compressed post-ocular plate.

The antennæ are completely exposed, from the base of the second joint of the peduncle.

The chelipeds in the male are massive, and in length are more than half again as long as the carapace and rostrum : all their joints, from

50

the ischium to the propodite, have one or more of their edges conspicuously and sharply cristiform, this being specially well marked in the
case of the long trigonal meropodite, which has all its edges sharply
phalanged, and in the case of the equally long slightly inflated palm,
which has razor-like edges. The fingers, which are not nearly half the
length of the palm, are acute, and have their cutting edges entire.

The 2nd-5th pairs of legs are slender, with cylindrical joints, the
2nd are nearly or quite equal in length to the chelipeds, the 3rd-5th
decrease gradually in size.

In an adult female, equal in size to the male above described, the
chelipeds are shorter than the 2nd pair of legs, and are similar in
general proportions to the other legs.

Colours in life : " Earth-colour with the chelipeds pink."

	Male.	Female (adult.)
Length of carapace (including rostrum)...	18 millim. ...	15·5 millim.
Greatest breadth of carapace ...	12·5 ,, ...	11·5 ,,
Length of cheliped	29 ,, ...	14 ,,
Greatest breadth of palm	4·5 ,, ...	1 ,,

Loc. Bay of Bengal, 193 and 210 fathoms.

The ova are large (diam. 1 millim.) and rather few in number.

In young males the chelipeds are of proportions intermediate
between those of the adult male and female.

Scyramathia globulifera, Wood-Mason.

Pugettia globulifera, Wood-Mason, Ann. Mag. Nat. Hist. March, 1891, p. 260.

Distinguished by the vertically erect ear-like hepatic spine, the
base of which forms a great polished bulla on either side of the
buccal frame, giving the animal, when viewed front end on, a bat-like
appearance.

The body and legs are downy, the legs being fringed with short
broad curly hairs.

The carapace, in which the cardiac region is broad and prominent
and not, as in *S. beauchampi*, narrow and sunken, has, besides the hepatic
spine already mentioned, the following marks :—

On the branchial regions, below and anteriorly, a sharp sinuous
human-ear-shaped crest ; above and posteriorly a spine ; and near the
middle line anteriorly an acumination. On the gastric region four faint
51

clevations. On the cardiac region, and also on the intestinal region, in the middle line, an acuminate eminence.

The rostrum consists of two divergent spines, about one-third the length of the rest of the carapace.

The eyes stand well out from beneath the pre-ocular spine, and are retractile against a small post-ocular tooth.

The other appendages closely resemble those of the preceding species; but the chelipeds, in the adult male, are shorter, being only equal in length to the carapace and rostrum, and the fingers have their cutting edges crenulate instead of smooth.

In females and in young males the chelipeds have the same relative proportions as in *Scyramathia beauchampi*.

	Male.	Female (adult).
Length of carapace (including rostrum)...	17 millim....	13 millim.
Greatest breadth of carapace... ...	10 ,, ...	7·5 ,,
Length of cheliped	18 ,, ...	9·5 ,,
Greatest breadth of palm	4 ,, ...	1·2 ,,

Loc. Andaman Sea, 130–240 fathoms.

Miers *Pugettia velutina* ('Challenger' Brachyura, p. 41, pl. vi. figs. 2, 2*a*, 2*b*) should, I think, be placed in this sub-genus—*Scyramathia*.

HYASTENUS, White.

Hyastenus, White, P. Z. S., 1847, p. 56.
Hyastenus, Miers, Journ. Linn. Soc., Zool., Vol. XIV. 1879, p. 658 (*et synon.*) ; and 'Challenger' Brachyura, p. 55.
Chorilia and *Lahainia,* Dana, U. S. Expl. Exp. Crust. I. pp. 91 and 92.

Carapace subpyriform, convex, either smooth or tuberculate, sometimes spiny. Supra-ocular eave very prominent, usually somewhat acuminately produced anteriorly : post-ocular spine, or lobe, large and excavated. The rostrum consists of two usually long slender divergent spines. Eye-stalks short, retractile against the post-ocular lobe, but never to the complete concealment of the cornea.

Basal antennal joint broad, its antero-external angle sometimes produced : the mobile portion of the antenna usually exposed to dorsal view.

Merus of the external maxillipeds as broad as, or broader than, the ischium, expanded at the antero-external angle, and bearing the palp at the antero-internal angle.

Chelipeds in the adult male enlarged : the second pair of trunk-legs usually very much longer than the 3rd 4th and 5th pairs. The abdomen in both sexes consists of seven distinct segments.

Key to the Indian species of Hyastenus.

	1. R o s t r a l spines at least as long as the carapace pro-per.	i. Rostral spines as long as the carapace, and nearly parallel in their proximal half : cara-pace indistinctly tubercu-lated .... [*H. sebæ.*]	
		ii. Rostral spines about twice as long as the carapace, and widely divergent from their origin : carapace with nu-merous tubercles, and with large cardiac, branchial and intestinal spines : a long ter-minal spine on the merus of of the second pair of trunk-legs .... *H. tenuicornis.*	
I. Denuded carapace with nume-rous tuber-cles, or spines, and erosions.	2. R o s t r a l spines not much more than half the length of the carapace pro-per.	i. Legs coarse, the mero-podites of all (includ-ing the chelipeds) nodular.	*a.* N u m e r o u s tubercles for-ming a cross on the gastric region : a me-dian trans-verse tuber-cle in the groove be-tween the gastric and c a r d i a c regions......... *H. pleione.*
			b. Gastric re-gion almost smooth : no tubercle be-tween the gastric and cardiac re-gions.......... *H. hilgendorfii.*
		ii. Legs slend-er, the me-ropodites smooth.	*a.* C a r a p a c e e l o n g a t e closely cover-ed with gra-nules and tu-bercles, with-out spines..... *H. oryx.*
			b. Conspicuous-ly large spines on the cardiac and branchial regions......... *H. gracilirostris.*

II. Denuded carapace smooth and polished, with a few large spines.

1. Carapace triangular, with a large epibranchial spine and at least one large sub-hepatic tubercle on either side.
  - i. A large intestinal and two large gastric spines in the middle line .... *H. spinosus.*
  - ii. No large intestinal spine: a single gastric tubercle in the middle line .... *H. diacanthus.*

2. Carapace elongate, with a small epibranchial tubercle, and with none of the sub-hepatic tubercles enlarged.
  - i. A pair of gastric tubercles in the middle line .... *H. aries.*
  - ii. Gastric region without tubercles.
    - a. An erect claw-like intestinal spine   *H. calvarius.*
    - b. No intestinal spine .... *H. planasius.*

Hyastenus pleione (Herbst).

Cancer pleione, Herbst, Krabben, III. iii. 52, taf. lviii. fig. 5.
Naxia pleione, Gerstaecker. Archiv. fur Naturgesch. XXII. 1856, p. 114, taf. v. figs. 1-2.
Hyastenus pleione, A. Milne-Edwards, Nouv. Archiv. du Mus. VIII. 1872, p. 250.
Hyastenus pleione, de Man, Archiv. fur Naturgesch. LIII. 1887, p. 225, taf. vii. fig. 3 ; and Journ. Linn Soc., Zool., Vol. XXII. 1888, p. 18.
Hyastenus pleione, Miers, 'Challenger' Brachyura, p. 56.
Hyastenus pleione, J. R. Henderson, Trans. Linn. Soc. (2) V. 1893, p. 343.

Carapace triangular, elegantly rounded behind, pubescent like the legs and rostrum, the regions well-defined, tuberculated as follows :— six tubercles disposed in a Y or cross on the gastric region, one in the groove between the gastric and the extremely prominent cardiac region, one in the middle of the intestinal region, and three in a line on the boundary of the hepatic and pterygostomian regions; on either branchial region are two longitudinal rows of tubercles, the upper row being the more distinct, but the last tubercle in the lower row being the largest, and forming a rather prominent epibranchial spine; finally on either side of the groove separating the cardiac and intestinal regions is a prominent tooth.

The rostrum consists of two slender divergent spines, which in the male are half the length of the carapace proper, but in the female are considerably less.

The basal antennal joint has its outer margin, anteriorly, bilobed.

The hairy trunk-legs have the upper surface somewhat uneven or actually nodular.

The chelipeds in the male are stouter than the other legs, and are as long as the carapace *plus* half the rostrum ; the fingers, which are hardly one half shorter than the short palm, are arched and meet only near their tips : in the female the chelipeds are rather more slender than the other legs, are only as long as the post-ocular portion of the carapace, and have nearly straight fingers that meet in the greater part of their extent.

The second pair of legs, in both sexes, are considerably longer than the chelipeds and than any of the three last pairs : the dactyli of all the ambulatory legs are stout, recurved, and serrated along the posterior margin.

In the Museum collection are numerous specimens of both sexes, from Ceylon and Mergui.

Hyastenus hilgendorfii, de Man.

Hyastenus hilgendorfii, de Man, Journ. Linn. Soc., Zool., Vol. XXII. 1888, p. 14, pl. i. figs. 3 and 4.

This species much resembles *H. pleione*, but is distinguished by the following constant characters :—the carapace is but faintly tuberculated, and, in particular, there is no tubercle between the gastric and cardiac regions : the dactyli of the ambulatory legs are very strongly toothed, instead of merely serrated, along the posterior margin : in the male the rostrum is nearly two-thirds the length of the carapace, and the chelipeds are as long as the carapace and rostrum combined, and nearly as long as the second pair of trunk-legs,—this being largely due to the increased length of the palm.

Carapace subpyriform, and, like the rostrum and legs, pubescent ; the regions moderately well-defined.

The gastric region is either quite smooth, or presents three faint elevations disposed in a triangle base forwards. There is a small tubercle near the middle of the intestinal region ; and a line of granulations along the boundary between the hepatic and pterygostomian regions, which line is continued backwards, along the side of the branchial region, to end at a distinct lateral epibranchial spine : there is also a more or less distinct line of granules on the dorsal aspect of the epibranchial region.

The rostrum consists of two divergent spines, the length of which in the male is nearly two-thirds that of the carapace proper, but is considerably less in the female. Basal antennal joint with the outer margin sinuously curved.

The trunk-legs have the surface somewhat uneven : the chelipeds in the male are much stouter than the other legs, and are as long as the
55

carapace and rostrum, the palm being nearly twice the length of the fingers, which are not much arched and meet in their distal half : in the female the chelipeds are rather slenderer than the other legs, and are equal to the postrostral portion of the carapace in length. The 2nd pair of legs are hardly longer than the (male) chelipeds, but are very much longer than the last three pairs : the dactyli in all are stout, recurved, and strongly toothed along the posterior margin.

Specimens are in the Museum collection from Ceylon, Ganjam, Mergui, the Nicobars, and the Straits of Malacca.

Hyastenus diacanthus (de Haan).

Pisa (Naxia) diacantha, de Haan, Faun. Japon. Crust., p. 96, pl. xxiv. fig. 1.
Naxia diacantha, Adams and White, ' Samarang ' Crust., p. 10.
Naxia diacantha, Stimpson, Proc. Acad. Nat. Sci. Philad. 1857, p. 218.
Naxia diacantha, Heller, ' Novara ' Crust., p. 3.
Hyastenus diacanthus, A. Milne-Edwards, Nouv. Archiv. du Mus. VIII. 1872, p. 250.
Naxia diacantha, Brocchi, Ann. Sci. Nat. (6) II. 1875, Art. 2, p. 94, pl. xix. figs. 172, 173 (male appendages).
Hyastenus diacanthus, Miers, Cat. Crust. New Zealand, p. 9 ; and P. Z. S., 1879, pp. 19 and 26 ; and Zoology H. M. S. ' Alert,' pp. 182 and 194 ; and ' Challenger ' Brachyura, p. 57.
Hyastenus diacanthus, Haswell, P. L. S., N. S. Wales, Vol. IV. 1879, p. 442 ; and Cat. Austral. Crust., p. 20.
Hyastenus diacanthus, de Man, Archiv. fur Naturgesch., LIII. 1887, p. 220.
Naxia diacantha, C. W. S. Aurivillius, Kongl. Sv. Vet. Akad. Handl. XXIII. 1888-89, No. 4, p. 51, pl. ii. fig. 5.
[*Hyastenus diacanthus,* Cano, Boll. Soc. Nat. Napol. III. 1889, p. 178.]
Hyastenus diacanthus, A. O. Walker, Journ. Linn. Soc., Zool., Vol. XX. 1890, p. 109.
Hyastenus diacanthus, Ortmann, Zool. Jahrb., Syst., etc., VII. 1893, p. 55 ; and Zool. Forsch. Austral. Malay. Archip., Jena., 1894, p. 42.
Hyastenus diacanthus, Mary Rathbun, Proc. U. S. Nat. Mus. Vol. XVI. 1893, p. 85.

Body and legs densely tomentose, often much encrusted with sponges, etc. Carapace pyriform, with the regions strongly convex, well-defined, and when denuded, smooth and polished : on the gastric region, in the middle line, there is an acuminate tubercle, on either pterygostomian region at least one large tooth, and near the hinder limit of either branchial region a horizontally projecting lateral epibranchial spine.

The rostrum consists of two more or less divergent horns, the length of which in the adult male is from half to nearly two-thirds that of the carapace proper, but in the female is less. The basal antennal joint is much inflated behind and constricted in front.

The chelipeds in the male are stouter than any of the other legs, and are equal in length to the carapace *plus* half the rostrum ; the fingers, which are arched and meet in rather less than their distal half, are nearly as long as the short inflated palm. In the female and young male the chelipeds are rather more slender than any of the other legs, and in length are equal to the post-ocular portion of the carapace ; and the fingers, which are almost straight, meet in the greater part of their extent. The second pair of trunk-legs are nearly twice the length of the (male) chelipeds, and are far longer than any of the last three pairs : the recurved and densely tomentose dactyli have the posterior margin almost smooth.

Besides specimens from the Australian and Chinese Seas, the Museum possesses specimens from Ceylon, Orissa, Tavoy, and the Andamans.

Hyastenus spinosus, A. Milne-Edwards.

Hyastenus spinosus, A. Milne-Edwards, Nouv. Archiv. du Mus., VIII. 1872, p. 250.
Hyastenus spinosus, Miers, 'Challenger' Brachyura, p. 56.

This species differs from *H. diacanthus* only in the following particulars :—the body and limbs are less densely tomentose ; the gastric region, instead of a single acuminate tubercle, has two strong spines in the middle line ; there is a stout spine, in the middle line, close to the posterior border of the carapace ; the lateral epibranchial spines are larger.

These differences are constant in a large series of specimens from different parts of the sea-coast of India : but in two specimens which seem referable to this species the gastric region is quite smooth, though abnormally convex.

Hyastenus aries (Latr.)

[*Pisa aries,* Latr. Encyc. X. p. 140].
Chorinus aries, Milne-Edwards, Hist. Nat. Crust. I. 315.
Hyastenus aries, A. Milne-Edwards, Nouv. Archiv. du Mus., VIII. 1872, p. 250.
Chorinus aries, Hilgendorf, MB. Ak. Wiss. Berl. 1878, p 786.
Chorinus aries, E. Nauck, Zeits. Wiss. Zool. XXXIV. 1880, p. 41 (gastric teeth).
Hyastenus aries, Miers, 'Challenger' Brachyura, p. 56.

Very closely resembling *H. spinosus,* from which it differs only in the following particulars—adult males of nearly equal size being compared :—(1) the rostral horns, instead of being long cylindrical divergent and down-curved only at tip, are short (being only one-third the length of the carapace proper in the male, and only about one-fourth

57

in the female), somewhat compressed horizontally, almost parallel or even a little incurved, and perceptibly though very slightly deflexed from the base ; (2) the carapace is much more convex and swollen, with the lateral epibranchial and the median posterior spines much smaller ; (3) the chelipeds have the palm less enlarged, and the fingers nearly straight, instead of arched ; (4) the anterior angle of the supra-orbital cave, instead of being sharply produced, is obtuse.

The Museum possesses specimens from the Orissa Coast and Gulf of Martaban, and also from the Straits of Malacca.

Hyastenus planasius, Ad. & White.

Pisa planasia, Adams and White, ' Samarang ' Crust., p. 9, pl. ii. figs. 4 and 5.
Hyastenus planasius, A. Milne-Edwards, Nouv. Archiv. du Mus., VIII. 1872, p. 250.
Hyastenus (Chorilia) planasius, Miers, Zoology H. M. S. ' Alert,' pp. 182 and 196 ; and ' Challenger ' Brachyura, p. 57.
Hyastenus planasius, Walker, Journ. Linn. Soc. Zool. Vol. XX. p. 109.

Carapace elongate-ovate, its surface smooth and polished anteriorly, finely granulose posteriorly, and with scattered tufts of hairs : a small eminence in the middle of the gastric region, and a small lateral epibranchial spinule, in front of which latter there may be a line of granules : lateral margin with three spinules anteriorly, two of which are on the pterygostomian region.

The rostrum is formed by two parallel spines, the tips of which are somewhat incurved, and the length of which is about one-sixth that of the carapace proper. The supra-ocular margin is, as usual, very prominent, and has its anterior angle somewhat produced. The antero-external angle of the basal antennal joint forms a distinct tooth visible from above. The legs are tomentose with additional long scattered setæ : the second pair (1st ambulatory legs) are, as usual, markedly the longest, being half again as long as the carapace and rostrum : the dactyli are short, stout, recurved, and serrated posteriorly. The chelipeds are described by Adams and White as follows :—" small, slender, equal in size, covered with scattered long stout hairs ; the third joint subcylindrical, curved inwards and enlarged anteriorly ; fourth joint short, rounded, and curved, with two small tubercles on the outer and upper surface ; fifth joint rather slender, sub-cylindrical, laterally compressed ; claws slightly gaping in the middle, curved inwards, and finely denticulated." As, however, the male specimen figured does not seem to be adult, these characters are perhaps changeable with age.

In the Museum collection are a young male and female from Ganjam and Arrakan.

Hyastenus calvarius, n. sp.

This species—females alone being available for comparison—differs from *H. planasius* chiefly in the following characters :—(1) there is an erect claw-like spine on the posterior border of the carapace in the middle line ; (2) the spines of the rostrum are straight, divergent, and about half the length of the carapace ; (3) the dactyli are longer and slenderer.

Three females—two of which are laden with eggs—from the Andamans. The larger egg-laden female measures 14 millim. from the tip of the rostrum to the posterior border of the carapace.

Description of the female.

Carapace elongate-ovate, with the surface, when denuded of scattered setæ, smooth and polished : the gastric region is very convex : the only armature of the carapace is (1) a large erect claw-like spine near the posterior border in the middle line, (2) a small lateral epibranchial spinule on either side, and (3) two or three granules along the antero-lateral border in the pterygostomian region. The rostrum is formed of two straight divergent spines, the length of which is about half that of the carapace proper. The antero-lateral angle of the prominent supra-ocular eave is sharp ; and that of the basal antennal joint is produced to form a spine which is plainly visible from above.

The legs are more or less fringed with stout club-shaped hairs : the second pair are, as usual, the longest : the dactyli are long and slender, and are recurved, with the posterior margin serrate. The chelipeds are slender, and the fingers meet in the greater part of their extent.

Hyastenus sebæ, White.

Seba, Thesaurus, III. xviii. 12.
Hyastenus sebæ, White, P. Z S., 1847, p. 57 ; and Ann. Mag. Nat. Hist., Vol. XX. 1847, p. 61 ; and 'Samarang' Crustacea, p. 11.
Hyastenus sebæ, A. Milne-Edwards, Nouv. Archiv. du Mus., VIII. 1872, p. 249.
Hyastenus sebæ, de Man, Archiv. fur Naturgesch., LIII. 1887, p. 223.
Hyastenus sebæ, Miers, 'Challenger' Brachyura, p. 56.
Hyastenus sebæ, Ortmann, Zool. Forsch. Austral. Malay. Archip. Jena, 1894, p. 42.

Carapace very elongate-triangular, its surface eroded and sculptured, but without distinct tubercles or spines. The rostral spines, which are equal in length to the carapace, are paralled in their proximal half. The chelipeds in the male are equal in length to the carapace *plus* one-third of the rostrum : their merus is not much stouter than that of the next pair of legs, but the palm is broadened and somewhat inflated : the fingers, which are hardly more than half the length of

59

the palm, are arched, and meet only at the tip. The other legs are slender, the second pair being much longer than the last three pairs and longer than the chelipeds.

The Museum possesses a specimen from Mauritius, which I have included here for the sake of comparison.

Hyastenus oryx, A. Milne-Edwards.

Hyastenus oryx, A. Milne-Edwards, Nouv. Archiv. du Mus., VIII. 1872, p. 250, pl. xiv. fig. 1.
Hyastenus oryx, Haswell, Proc. Linn. Soc., N S. Wales, Vol. IV. 1879, p. 442 ; and Cat. Austral. Crust., p 20.
Hyastenus (Chorilia) oryx, Miers, Zool. H. M. S. 'Alert,' pp. 182 and 195, 517 and 522 ; and 'Challenger' Brachyura, p. 58.
Hyastenus oryx, de Man, Archiv. fur Naturgesch., LIII. 1887, p. 224, taf. vii. fig. 2.
Hyastenus oryx, C. W. S. Aurivillius, Kongl. Sv. Vet. Akad. Handl. XXIII. 1888-89, No. 4, p. 50, pl. iv. fig. 4.
Hyastenus oryx, A. O. Walker, Journ. Linn. Soc., Zool., Vol. XX. 1890, p. 109.

Carapace pyriform, little setose, crisply and rather closely tuberculated, but without any spines, the tubercles on the gastric region being disposed in the form of a cross or anchor. The rostrum consists of two slender horns, which in the male are about half the length of the carapace proper, and strongly resemble the horns of an Oryx in miniature : in the female they are not one-third the length of the carapace, and are nearly parallel.

The supra-ocular eave is sharply angled, but not produced, anteriorly. The basal antennal joint is sharply toothed at the antero-external angle.

The chelipeds in the male are as long as the carapace *plus* two-thirds of the rostrum , their merus is slender, but the palms are broadened and inflated ; and the fingers, which are from half to two-thirds the length of the palm, are arched, and meet only at the tip. In the female the chelipeds are considerably shorter than the post-ocular portion of the carapace, and are rather more slender than the ambulatory legs, the fingers being but little arched, and little separated when clenched.

The ambulatory legs are slender, with slender almost smooth actyli : the first pair, which are considerably the longest, are about one-fourth longer than the carapace and rostrum.

This, like *Hyastenus calvarius*, is a small species, an egg-laden female of average size measuring only 14 millim. from the tip of the trum to the posterior border of the carapace. It is a common species at the Andamans, and has also been taken off Ceylon at 34 fathoms.

Hyastenus gracilirostris, Miers,

Hyastenus gracilirostris, Miers, Ann. Mag. Nat. Hist., Vol. IV. 1879, p. 12, pl. iv. fig. 7 ; and 'Challenger' Brachyura, p. 56.

Carapace subpyriform, hardly at all setose, with numerous sharp tubercles and spinules. Specially noticeable are three spinules, longitudinally arranged in the middle line, on the gastric region, a strong conical spine on the cardiac region, a sharp tubercle on the posterior margin, and two spines on each of the branchial regions, of which the larger occupies the usual position of the lateral epibranchial spine.

The rostrum, which does not vary according to sex, consists of two slender divergent spines, the length of which is about one-third that of the carapace. The post-ocular lobe projects very strongly, and the supra-ocular eave has both the anterior and the posterior angle pronounced. The basal antennal joint has a well-marked tubercle or blunt spine at its antero-external angle.

The chelipeds in the male are equal in length to the post-rostral portion of the carapace, and have a few small granules on the merus carpus and upper edge of the palm ; the merus is slender, but the palm is broadened and is not much longer than the fingers, which are arched and meet only at the tip. In the female the chelipeds are rather shorter than the post-ocular portion of the carapace, are very slender, and have nearly straight fingers.

The ambulatory legs are slender, with slender smooth-edged dactyli : the first pair are, as usual, much the longest.

This also is a small· species, and egg-laden female of average size being only 10 millim. in length.

In the Museum collection are specimens from the Madras coast.

Hyastenus tenuicornis, Pocock.

Hyastenus tenuicornis, Pocock, Ann. Mag. Nat. Hist., Vol. V. 1890, p. 76.

Distinguished by the enormous length of the rostral spines, and by the curious form—described below—of the supra-ocular eave and post-ocular lobe.

Carapace subpyriform, somewhat depressed, with the regions well-defined ; its surface with many long scattered setæ, and with numerous granules and some large spines. Specially noticeable are five or seven granules, arranged in the form of a cross, on the gastric region ; two huge acuminate tubercles, in the middle line, posteriorly; and three spines on either branchial region, the hindmost and lowermost of which is of great size.

The rostrum consists of two slender, exceedingly divergent spines,

61

the length of which in the male is about twice, in the female about once and a fifth, that of the carapace.

The post-ocular lobe is unique is form : it is very prominent, and has a stout pedicle and a compressed crown, the angles of which are produced. The supra-ocular eave is also unique: it also is very prominent, and has its antero-external angle produced forwards and upwards, and its postero-external angle produced backwards towards the post-ocular lobe. The basal antennal joint is deeply grooved longitudinally : its antero-external angle forms a strong spine visible from above, and its outer edge bears two distinct teeth which stretch towards the supra-ocular and post-ocular spines respectively. All the trunk-legs are very slender: the first two pairs have a strong spine on the far end of the upper border of the merus, but this in the last three pairs is represented by a small tubercle. The chelipeds, even in the male, are slender throughout, and have long slender fluted palms which are three times the length of the fingers : the latter, though denticulated throughout and but little arched, meet, in the male, only in their distal half.

The first pair of ambulatory legs are, as usual, much the longest : in all the dactyli are long and slender, but have the posterior edge sharply serrated.

This also is a small species, an egg-laden female of average size measuring only 17 millim., more than half of which is rostrum.

Off Cheduba (Arakan coast) 7 fathoms : off Ceylon 30–34 fathoms.

Dr. Henderson (Tr. Linn. Soc., Zool., 1893, p. 344) also includes in the Indian Fauna, but with some doubt, the two following species :—

1. *Hyastenus convexus*, Miers Zool., H. M. S. 'Alert,' p. 196; pl. xviii. fig. B. (*N. E. Australia ; Penang.*).

2. *Hyastenus brockii*, de Man, Archiv. fur Naturgesch. LIII., 1887, p. 221, taf. vii. fig. 1. (*Amboina*).

As Dr. Henderson seems to be not quite sure of his identification, and as we have no specimens in the Museum collection, I have not noticed these two species at length.

NAXIA, Edw., Miers.

Naxia, Milne-Edwards, Hist Nat. Crust. I. 313.
Naxia, de Haan. Faun, Japon. Crust , p. 84.
Naxia, Miers, Journ. Linn. Soc., Zool., Vol XIV. 1879, p. 658 (*et synon.* Naxioides, *A. M. Edw. and* Podopisa *Hilgendorf) ;* and 'Challenger' Brachyura, p. 59.

Carapace subpyriform, moderately convex, rounded behind, and armed with spines or tubercles on the dorsal surface. Spines of the rostrum well developed, subcylindrical, parallel or divergent, and bearing on the inner margin, near to the extremity, a small accessory spine or spinule. Abdomen (in the male) distinctly seven-jointed; in the female some of the segments may be coalescent. Eyes small, supra-ocular eave very prominent, its antero-external angle sometimes produced to a spine : post-ocular lobe also very prominent, its edge unequally bi- or tri-lobed. Antennæ with the basal joint enlarged, with a spine or tubercle at the antero-lateral angle, and sometimes with another on the outer margin ; the flagellum either exposed, or partially concealed in a dorsal view by the rostral spines. Merus of the external maxillipeds distally truncated, with the antero-external angle little, if at all, produced, and the antero-internal angle emarginate. Chelipeds (in the male) slender and moderately developed, palm usually somewhat elongated, fingers denticulated near the distal extremity, and having between them when closed a small hiatus at the base. Ambulatory legs slender and somewhat elongated, the first pair much the longest, with the joints subcylindrical ; dactyli nearly straight.

Key to the Indian species of Naxia.

I. Armature of the carapace consisting almost entirely of large clean-cut spines .... *N. hystrix.*

1. Spines of the rostrum parallel to near the tip: supra-ocular spine obsolete : meropodites of the trunk-legs without a terminal spine...... *N. hirta.*

II. Armature of the carapace consisting chiefly of tubercles, among which there are sometimes a few coarse spines.

2. Spines of the rostrum divergent from the base: supra-ocular spine present: meropodites of some of the trunk-legs with a large terminal spine.

i. Spines of the rostrum considerably more than half the length of the carapace: supra-ocular spine very large and acute : meropodites of all the trunk-legs with a terminal spine : palms long and slender.

a. Rostral spines widely divergent: no large spines on the branchial or intestinal regions ...... *N. taurus.*

b. Rostral spines moderately divergent : several large spines on the branchial regions and in the middle line of the carapace *N. cerastes.*

ii. Spines of the rostrum considerably less than half the length of the carapace : supra-ocular spine blunt : meropodites of the last three pairs of trunk-legs unarmed : palms short and inflated .... *N. investigatoris.*

Naxia investigatoris, n. sp. Pl. IV. fig. 3.

Distinguished from all other Indian species by the form of the male chelipeds, of which the palm, instead of being long and slender, is short and broadly inflated.

Carapace subpyriform, with all the regions well-defined, and the whole surface, from the base of the rostral spines, sharply tubercular.

The rostral spines in the male and sometimes in the female are hardly one-third the length of the carapace proper, and are divergent, with the accessory spine in the middle of the distal half : often, but not always, in the female they are less than one-fourth the length of the carapace, are little divergent, and bear the accessory spinule near the tip. The antero-external angle of the prominent supra-ocular eave is surmounted by a blunt spine : the basal antennal joint has a similar spine at its antero-external angle, and another near the middle of its outer border.

The chelipeds are granular, and their meropodite has a small spinule at the distal end of its upper border : in the male they are a little longer than the carapace, the palm is short—less than twice the length of the fingers—inflated, and enlarged from behind forwards, and the fingers are strongly arched and meet only at the tip : in the female they are only as long as the post-rostral portion of the carapace, are slender throughout, and have nearly straight fingers. The 2nd pair of trunk-legs (1st pair of ambulatory legs) are $2\frac{1}{2}$ times the length of the carapace, and have the meropodite armed with a strong spine at the distal end of its upper border, and the dactylus of remarkable length, nearly equal to the propodite : the other legs are much shorter, and have the spine replaced by a small tubercle, their dactylus being of ordinary length.

Colours in spirit, pale ochre.

Loc. Andamans ; and off Ceylon, 34 fathoms.

	Male.	Ovigerous Female.
Length of carapace and rostrum ...	19 millim.	17 millim.
Greatest breadth of carapace ...	10·5 ,,	10 ,,
Length of chelipeds...	23 ,,	14 ,,
Length of 2nd pair of legs... ...	41 ,,	36 ,,

Naxia hirta, A. Milne-Edwards.

Naxioides hirta, A. Milne-Edwards, Ann. Soc. Ent. Fr. (4) V. 1865, p. 143, pl. iv. fig. 1.

Podopisa petersii, Hilgendorf, MB. Ak. Berl., 1878, p. 785, taf. i. figs. 1–5.

Naxia petersii, Miers, Zoology of H. M. S. 'Alert,' p. 523.

Naxia hirta, Miers, 'Challenger' Brachyura, p. 61.
Naxia petersii, de Man, Journ. Linn. Soc., Zool., Vol. XXII. 1888, p. 19.
Naxia hirta, Pocock, Ann. Mag. Nat. Hist., Vol. V. 1890, p. 79.
Naxia hirta, Henderson, Traus Linn. Soc., Zool. (2) V. 1893, p. 345.

Carapace pyriform, with the regions well-defined and the surface from the base of the rostral spines unevenly granular and tubercular. From the rough surface there stand out (1) at least two good sized spines on either branchial region, (2) a sharp unciform tubercle close to the posterior border near the middle line, and (3) a stout nipple-shaped tubercle near the middle of the pterygostomian region.

The rostral spines, which in both sexes are close together and parallel in more than half their extent, are from one-third (male) to two-sevenths (female) the length of the carapace proper; from the point of origin of the accessory spines, which are situated at the end of the parallel portion, they are elegantly divergent.

The prominent supra-ocular eave has the antero-external angle slightly upturned. The basal antennal joint has a stout spine anteriorly, and a coarse tooth in the middle of its outer border.

The chelipeds are smooth, and are slender in both sexes, but most so in the female: in the male they are equal in length to the post-rostral, in the female to the post-ocular portion of the carapace: the palms are slender and sub-cylindrical, and are twice the length of the fingers, which latter are hardly arched, and are therefore but slightly separated at the base when clenched.

All the ambulatory legs are slender and smooth, and the first pair are considerably the longest, being nearly twice the length of the carapace and rostrum, the dactylus not being abnormally elongate. The body and legs are covered with a short fine down, and the colour in spirit is usually mottled reddish and yellow.

In the Museum collection are specimens from the Andamans and from Ceylon.

Naxia taurus, Pocock.

Naxia taurus, Pocock, Ann. Mag. Nat. Hist. Vol. V. 1890, pp. 77 and 79.
Naxia taurus, Henderson, Trans. Linn. Soc., Zool. (2) V. 1893, p. 346.

Distinguished by the very long and widely divergent rostral spines.

Carapace pyriform, with the regions well-defined, and the surface, from the base of the rostral spines, unevenly granular and tubercular beneath tufts of hair. Among the tubercles three on the gastric region in the middle line, three in a triangle on the intestinal region, and three on either branchial region attract attention.
65

The rostral spines, which are considerably more than half the carapace in length, are widely divergent—the distance between their tips being more than three-quarters of their length : the accessory spine is situated not far in front of the middle.

The prominent supra-ocular eave has a strong sharp spine, and there is an even stronger and sharper spine at the antero-external angle of the basal antennal joint, as well as a prominent tooth near the middle of the outer border of this joint.

The chelipeds are granular : in the male they are equal in length to the carapace *plus* four-fifths of the rostrum, and, though slender, are considerably stouter than the other legs, especially as to the palm, which is more than twice the length of the fingers—the fingers being but little curved and therefore but little separated when closed : the meropodite has a strong sharp spine at the distal end of its upper border.

The ambulatory legs are slender : the meropodites of all but the last pair are armed as to the distal end of the upper border with a spine, which is of conspicuous size in the case of the first pair. The first pair are markedly the longest, being nearly twice the length of the carapace measured with the long rostrum, and have the dactylus extremely long—nearly equal in length to the propodite.

A single male specimen occurs in the collection, having been dredged off the Andamans in 36 fathoms.

Naxia cerastes, Ortmann.

Naxia cerastes, Ortmann, in Semon, Zool. Forschungreisen Austral. und Malay. Archipel., Crust., p. 43, taf. iii. fig. 4.

This species appears to be very similar to *Naxia taurus*, with which it may, perhaps, even be identical. It differs from *Naxia taurus*, comparing specimens of the same size and sex, in the following unimportant particulars :—(1) the rostral spines are less divergent ; (2) the carapace, in addition to the granules and tubercles, is armed with several large spines, of which three on either branchial region and one on the intestinal region are of conspicuous size, while several in the middle line on the gastric and cardiac regions are hardly smaller.

In the collection are a perfect male and female from the Andamans.

Naxia hystrix, Miers.

Naxia hystrix, Miers, ‘Challenger’ Brachyura, p. 60, pl. vi. fig. 4.
Naxia hystrix, R. I. Pocock, Ann. Mag. Nat. Hist., Vol. V. 1890, p. 79.
Naxia hystrix, Ortmann, Zool. Forsch. in Austral. und Malay. Archipel., Crust., p. 43.

Body closely beset with short knobbed hairs, among which longer setæ are interspersed.

Carapace subpyriform, armed with numerous long sharp spines as follows:—four, arranged in a triangle base forwards, on the gastric region; one on the cardiac, and one (very large) on the intestinal region; one on either hepatic region; two or three on either pterygostomian region; and, finally, on either branchial region three dorsal and three lateral: between these large spines some spinules and sharp granules are interspersed. In the male there is a pair of strong spines on the sternum between the chelipeds; and each abdominal tergum has a strong median spine: in the female five parallel rows of spines are found on the ventral surface, three of which belong to the abdominal terga, and one on either side to the sternum.

The rostral spines are short (about one-fifth the length of the carapace in the male, and rather less in the female), and divergent: the accessory spinule is found on their inner margin near the tip.

The basal antennal joint has a sharp spine at its antero-external angle, and a tooth near the middle of its outer margin. The antero-external angle of the prominent supra-ocular eave is surmounted by a sharp spine.

The chelipeds in the female and young male are rather more slender than the other legs, and are a little longer than the carapace and rostrum: the palms are slender and subcylindrical, and are nearly three times as long as the fingers, which are nearly straight and apposable throughout. The ambulatory legs are slender, and have very long slender dactyli: the first pair, which are much the longest, are nearly three times as long as the carapace and rostrum.

In the Museum collection are specimens from the Andaman Sea down to 40 fathoms.

CHORILIBINIA, Lockington, Miers.

Chorilibinia, Lockington, Proc. Ac. Nat. Sci. Calif., Vol. VII. 1876, p. 69.
Chorilibinia, Miers, Journ. Linn. Soc., Zool., Vol. XIV. 1879, p. 45; and 'Challenger' Brachyura, p. 45.
Chlorolibinia, Haswell, Cat. Austral. Crust., p. 17.

Carapace broadly subpyriform, spinose. Rostrum formed of two spines which are coalescent in their basal half. The commencing orbit, which does not afford much concealment to the fully retracted eye, is formed by a little-prominent supra-ocular eave, and a cupped (and isolated) post-ocular tooth. The basal antennal joint is broad, has its antero-external angle more or less produced, and has also a lobe on its

67

outer margin, near the base. Merus of the external maxillipeds as broad as the ischium, and with the antero-external angle produced.

Chelipeds slender ; ambulatory legs very long and slender. Abdomen of the male consisting of seven distinct segments.

Chorilibinia andamanica, n. sp. Plate V. figs. 2, 2a.

Distinguished from *Chorilibinia gracilipes*, Miers (Ann. Mag. Nat. Hist. Vol. IV. 1879, p. 7, pl. iv. fig. 4), (1) by the much less divergent rostral spines ; (2) by the pair of great spines—one pointing forwards, the other backwards—on the cardiac region ; (3) by the much slenderer chelipeds.

Carapace broadly subpyriform, with (1) a median line of tubercles and spines increasing in size from before backwards, four of the spines— namely one on the after limit of the gastric region, two on the cardiac region, and one near the posterior border—being conspicuously large ; and with (2) on either side a supra-marginal line of spines as follows— a tooth at the angle of the buccal frame, a large hepatic spine pointing downwards, and four branchial spines, the last of which directed obliquely backwards is much the largest. Besides these large spines there are numerous, symmetrically disposed, sharp granules. The rostrum, which measured from the anterior border of the orbit is about one-third the length of the carapace proper, ends in two very slightly divergent spines.

The eyes are short and thick ; and the orbit is formed by a moderately prominent supra-ocular eave separated by a narrow interval from a broad isolated post-ocular pocket.

The basal antennal joint is moderately broad, and bears two teeth, one at the antero-lateral angle, the other at the base—the latter inclining towards the post-ocular pocket.

The external maxillipeds completely close the buccal frame, the merus being as broad as the ischium.

The chelipeds are not stouter than the legs, and are but little longer than the carapace (rostrum included) : the next pair of legs are considerably more than three times, and the third pair are about three times, this length ; while the 4th and 5th pairs are very short.

The abdominal segments from the third to the sixth inclusive, are coalescent.

The sternum between the chelipeds carries a pair of very strong sharp teeth.

Loc. Andamans.

EGERIA, Leach.

Egeria, Leach, Zool. Miscell. Vol. II. p. 39.
Egeria, Milne-Edwards, Hist. Nat. Crust. I. 290.
Egeria, Miers, Journ. Linn. Soc., Zool., Vol. XIV. 1879, p. 654; and 'Challenger' Brachyura, p. 44.

Carapace subpyriform, nearly as broad as long, convex and tuber-culated. The rostrum consists of two vertically compressed spines of no great length, which are fused together in half or more of their extent. The eyes are short. The commencing orbits are formed by a supra-ocular eave and a post-ocular tooth, the interval between this tooth and the supra-ocular eave above, and between it and the basal antennal joint below, being partly closed in each case by a spine. The basal antennal joint is truncate-triangular; its antero-external angle is produced, and there is a second spine behind the middle of the outer border: the mobile portion of the antenna is visible from above on either side of the rostrum. The merus of the external maxillipeds is as broad as the ischium. Chelipeds in the adult male considerably longer than the carapace and rostrum, and having the palms inflated. Chelipeds in the female very slender throughout, and a little longer than the carapace and rostrum. Ambulatory legs extremely long and slender, the first pair being about six times the length of the carapace and rostrum: the dactylus in all is remarkably long. Abdomen of male seven-jointed: of female five-jointed.

Egeria arachnoides (Rumph), Edw.

Egeria arachnoides, Rumph, pl. viii. fig. 4; [and Latreille, Encyc. Pl. 281, fig. 1 ;] and Milne-Edwards, Hist. Nat. Crust., I. 291; and Neumann, Syst. Uebers., 1878, p. 19; and Haswell, P. L. S., N. S Wales, IV. 1879, p. 439, and Cat. Austr. Crust., p. 11; and Miers Zool. Alert, pp. 182 and 191, and 'Challenger' Brachyura, p. 44; and C. W. S. Aurivillius, Kongl. Sv. Vet. Ak. Handl., XXIII. 1888-89, No. 4, p. 44; and Ortmann, Zool. Jahrb. Syst. etc., VII. 1893, p 48; and J. R. Henderson, Trans. Linn. Soc. Zool. (2) V. 1893, p. 343.

Cancer longipes, Herbst, Krabben, I. ii. 231, pl. xvi. fig. 93; and Fabricius Syst. Ent. ii. 466.

Inachus longipes, Fabr. Suppl., p. 358.

Macropus longipes, Latr. Hist. Nat. Crust. VI. 111.

Leptopus longipes, Lamk. Hist. An. Sansvert. V. 235; and Desmarest Consid. Crust. p. 159; [and Guérin, Icon. Reg. An. Crust., pl. x. fig. 3]; and Cuvier, Regne An. Crust., pl. xxxiv. fig. 1; and Adams and White, 'Samarang' Crust., p. 7; and Stimpson, Proc. Ac. Nat. Sci., Philad., 1857, p. 216; and A. O. Walker, Journ. Linn. Soc. Zool., XX. p. 109; and M. J. Rathbun, P. U. S. N. M., XVI. 1893, p. 95.

Egeria indica, Leach, Zool. Miscel. II. pl. lxxiii; and Desmarest, Consid. Crust., p. 157, pl. xxvi. fig. 2; and Milne-Edwards, Hist Nat. Crust. I. 292; and Adams and White, 'Samarang' Crust., p. 6; and E. Nauck, Zeits. Wiss. Zool. XXXIV. 1880, p. 41 (gastric teeth).

Egeria herbstii, Milne-Edwards, Hist Nat. Crust. I. 292; and Heller, 'Novara' Crust., p. 4; and Haswell, P. L. S., N. S. Wales, IV. 1879, p. 439, and Cat. Austr. Crust., p. 12.

Our large series of perfect specimens fully supports Mr. Miers' conclusion that all the hitherto described species of *Egeria* may be regarded as identical with the species rather poorly figured in Rumph's *Amboinische Rariteitkamer.*

Carapace subpyriform, or, rostrum excluded, subcircular, its breadth being equal to its length behind the base of the eye-stalks : the regions are distinctly delimited, and the surface is uneven and armed with some symmetrically disposed spines and spinules of which the six following are very conspicuously large, namely :—in the middle line, one on the cardiac and one on the intestinal region, and, on either side, a subhepatic and a lateral epibranchial : besides these there is (1) a conspicuous set of spinules arranged in the form of a T on the gastric region — the last in the vertical limb of the T being a distinct spine ; and (2) two series of distant spinules on either branchial region.

The rostrum varies somewhat : it is always short, and typically, consists of two vertically compressed spines which are fused in rather more than half their extent and have the tips slightly divergent : but sometimes the fusion is more extensive, or the tips are broken, and the rostrum then has the form of an emarginate stump. The supra-ocular eave is surmounted by a small sharp tooth anteriorly.

The chelipeds in the adult male are more than half again as long as the carapace and rostrum : the merus is a little enlarged distally, and the palm is inflated and distally enlarged : the fingers, which are half the length of the palm, are slightly separated at the base when clenched.

The chelipeds in the female are only one-fourth longer than the carapace and rostrum, and are the slenderest of all the trunk-legs.

The first pair of ambulatory legs are at least six times the length of the carapace and rostrum, rather more than a third of their extent being formed by the dactylus : the other legs gradually decrease in length to the fourth and last, which are about $2\frac{1}{3}$ times the length of the carapace and rostrum. The joints in all are very slender, cylindrical, and except for a spine at the distal end of the upper border of the merus, quite smooth.

Conspicuous on the sternum of the male is a pair of large teeth, placed between the front legs.

The body and legs are usually covered with an excessively short fine down : the legs are often banded, sometimes very distinctly, with dull red.

Egeria investigatoris, n. sp.

This species closely resembles *Egeria arachnoides,* adult males being compared, but differs in the following particulars : — (1) the carapace is more nearly circular, owing to the greater convexity of the hepatic and pterygostomian regions ; (2) the spines on the carapace, although almost the same in arrangement, are markedly larger : (3) the sternum has a transverse group of spines on every segment ; (4) every abdominal tergum except the last has a large median spine ; (5) the hiatus between the post-ocular tooth and the basal antennal joint is scarcely affected by a small denticle ; (6) the chelipeds in the adult male are $2\frac{1}{3}$ times the length of the carapace, and have the palm long, very slender, and cylindrical, and the fingers sharply and evenly denticulated all along their apposable edge.

The legs are in fragments, but the joints that remain are extremely long and slender.

Length of carapace and rostrum ... $24 + 5 = 29$ millim.
Breadth of carapace ... ... 24 „
Length of male chelipeds ... ... 65·5 „

Loc. Off Ceylon, 32 fathoms.

Doclea, Leach.

Doclea, Leach, Zool. Miscell., Vol. II. p. 41.
Doclea, Milne-Edwards, Hist. Nat. Crust. I. 292.
Doclea, Miers, Journ. Linn. Soc., Zool., Vol. XIV. 1879, p. 652.

Body and appendages tomentose, usually very densely so.

Carapace circular, armed at the sides, and often on the dorsal surface also, with a few spines.

The rostrum consists of two vertically compressed spines which are fused together in almost the whole of their extent and are usually short : it has hence, usually, the appearance of a short flat emarginate beak, hardly breaking the general outline of the carapace. (In one species — *Doclea tetraptera* — the rostrum is rather long).

The eyes are very small, and the commencing orbits are formed by an acute post-ocular tooth and a little-prominent supra-ocular eave. The antennæ are very short and inconspicuous — not reaching to the end of the short rostrum : the basal joint is short, broadly triangular, the apex forming a sharp tooth : the flagella are almost rudimentary.

The buccal frame is somewhat arched in front. The external maxillipeds have the merus rather broader than the ischium, the antero-external angle being slightly produced.

The chelipeds are short and slender in the female ; longer, stout, with an enlarged and inflated palm, in the adult male.

The abdomen consists of seven segments in the male, and of seven in the female of all except *D. muricata* and *hybrida*.

Key to the Indian species of Doclea.

I. Pterygostomian regions distinctly canaliculated fore and aft.

1. Rostrum elongate—one-fourth to two-fifths the length of the carapace proper, and with the points very widely divaricated: the last lateral and the median posterior spines of huge size.............. *D. tetraptera.*

2. Rostrum short—one-sixth the length of the carapace proper—and with no marked divergence of the tips.

    i. Two lateral spines on the branchial region: no median posterior spine.... .... *D. ovis.*

    ii Three lateral spines on the branchial region, the last being short: a short median posterior spine: no spines on the dorsum of the carapace....... *D. japonica.*

    iii. Three lateral spines on the branchial region, the last being, like the postero-median spine, long: a line of tubercles, two of which are usually produced to form spines, down the middle of the carapace........... *D. canalifera.*

II. Pterygostomian regions not canaliculated.

1. Carapace discoid: 2nd pair of trunk-legs three to four times the length of the carapace: a single series of tubercles or spines down the middle of the carapace... .... *D. gracilipes.*

2. Carapace globular: 2nd pair of trunk-legs hardly twice the length of the carapace: a short series of tubercles or spines on either branchial region parallel to a long mid-dorsal series of tubercles or spines

    i. Tubercles, not spines on the carapace.............. *D. hybrida.*

    ii. Spines not tubercles, on the carapace............ *D. muricata.*

Carcinological Fauna of India.   227

Doclea ovis (Herbst), Edw.

Cancer ovis, Herbst, Krabben, I. ii. 210, tab. xiii. fig. 82; and Fabricius, Syst.
Ent. II. 459.
Inachus ovis, Fabricius, Supplement, p. 355.
[Maia ovis, Bosc. I. 256]; and Latreille, Hist. Nat. Crust. VI. 100.
Doclea ovis, Milne-Edwards, Hist. Nat. Crust. I. 294.
Doclea ovis, Cuvier, Règne Animal, Crust., pl. xxxiii. fig. 2.
Doclea ovis, Adams and White, Zool. ' Samarang,' Crust., p. 7.
Doclea ovis, A. O. Walker, Journ. Linn. Soc., Zool., XX. 1890, p. 109.

Body and appendages, except the hands and the tips of the dactyli, covered with an extremely dense soft fur.

Beneath the fur the carapace is almost smooth, its surface being hardly broken by a median line of pimples on the gastric region; but its antero-lateral border, on each side, is armed with four sharp teeth of about equal size—one at the angle of the buccal frame; one, which has sometimes a tubercle at its base, on the sub-hepatic region; and two on the front part of the branchial region. The basal antennal joint has also the form of a tooth, and midway between it and the tooth at the outer angle of the buccal frame is another tooth. So that, including the pointed basal antennal joint, the antero-lateral margin of the carapace shows six teeth : there is no spine, though occasionally a trace of a tubercle, on the posterior border.

The rostrum hardly breaks the general subcircular outline of the carapace: it is cleft at the tip, and, measured at the level of the base of the post-ocular tooth, is broader than long.

The pterygostomian region is longitudinally grooved. The chelipeds in the old male are 1¼ times the length of the carapace and rostrum, and are enlarged, especially as to the palm, which is ¾ as broad as long, and is inflated on the inner side: the fingers also are stout and meet only in (about) the distal third. In the female the chelipeds are only about ¾ the length of the carapace and rostrum, and are throughout slenderer than the other legs. The 2nd pair of trunk-legs (first ambulatory legs) are from twice to 2¼ times the length of the carapace and rostrum.

The abdomen in both sexes consists of seven distinct segments, and the second segment in the female bears a large median elevation.

A common species in muddy waters in the vicinity of the mouths of the large rivers of India.

Doclea japonica, Ortmann.

Doclea japonica, Ortmann, Zool. Jahrb. Syst., &c., VII. 1893, p. 46, pl. iii. fig. 4.

The only differences between this species and Doclea ovis are (1)

73

that, instead of only two spines on the lateral border of the branchial region, there are *three*, the last being the largest and being placed rather higher up, (so that, including the tooth-like basal antennal joint, there are seven points on the antero-lateral border of the carapace) ; and (2) that there is a coarse spine, or blunt tooth, on the posterior border of the carapace.

I do not think that these differences are of more than varietal value ; for it is not uncommon in *Doclea ovis*, after careful denudation, to find traces of tubercles corresponding to the additional spines of *D. japonica.*

In the Museum collection are specimens from the mouth of the R. Hooghly.

Doclea canalifera, Stimpson.

Doclea canalifera, Stimpson, Proc. Acad. Nat. Sci., Philad., 1857, p. 217.

Body and appendages, except the fingers and dactylopodites, covered with a dense velvet-like tomentum. Carapace subcircular with a line of tubercles or spines down the middle line, namely, some minute tubercles (only visible on the denuded carapace), followed by a spine, on the gastric region ; a larger spine on the cardiac region ; and a much larger one still on the posterior border : the antero-lateral border is armed with four spines, the first bounding the outer edge of the pterygostomian canal, the last, which is rather larger than the spine of the posterior border, standing near the middle of the branchiostegal border : in addition, there is a small spine at the outer angle of the buccal frame, but no spine between this and the basal antennal joint ; and there is a line of extremely faint tubercles, only visible after complete denudation, stretching obliquely on either side from near the front towards the last epibranchial spine.

The rostrum, which is hardly longer than the breadth between the eyes, is sharply and deeply bifid at tip.

The pterygostomian region is longitudinally grooved. The chelipeds (in the young male) are slenderer than the next pair of legs, and are equal to the length of the carapace between the base of the rostrum and the base of the spine on the posterior border. The second pair of trunk-legs, which are the longest, are a little less than twice the length of the carapace and rostrum.

Abdomen of the male seven-jointed.

In the Museum are specimens from the mouth of the Hooghly and from the muddy estuarine coasts of Orissa and of Arakan.

74

Doclea gracilipes, Stimpson.

Doclea gracilipes, Stimpson, Proc. Ac. Nat. Sci., Philad., 1857, p. 216.
Doclea sp. De Man, Mergui Crust., Journ. Linn. Soc., Zool., XXII. 1888, p. 13.
Doclea andersoni, De Man, op. cit., tom. cit., p. 11, pl. i. fig. 1.

Carapace discoidal, covered, as are also the legs as far only as the end of their merus or carpus, with a short close fur.

Rostrum, measured from the posterior orbital border, sometimes as long as broad and about one-seventh the length of the carapace, sometimes twice as long as broad and about one-fourth the length of the carapace ; deeply cleft, the spines sometimes convergent, sometimes almost in contact throughout, sometimes slightly divergent.

Besides a line of four teeth, situated one at the end of the basal antennal joint, one at the angle of the buccal frame, and one behind each of these, the antero-lateral margin is armed with four acute curved claw-like spines, the posterior of which is typically two-fifths to one-third the breadth of the carapace in length, but may sometimes be only one-eighth the breadth of the carapace in length ; while the three anterior are typically about one-sixth the breadth of the carapace in length, but may sometimes be merely tubercles.

In the middle of the posterior border is a great spine as large as the last spine of the antero-lateral series.

In the middle line of the carapace is a series of tubercles and spines which are very variable in size : typically only two are prominent, and these have the form of upstanding spines, one in the gastric region, the other—much larger—in the cardiac region. Both of them, however, may be reduced to tubercles, while in front of them and also between them there may or may not be a line of tubercles.

Except for this median line of elevations, the dorsum of the denuded carapace is either smooth, or has only a line of extremely indistinct elevations passing on either side obliquely from near the front towards the great lateral epibranchial spine.

The chelipeds in the female are rather shorter than the carapace : in the male they are rather longer than the carapace, and in the adult male have the palms swollen.

The 2nd pair of trunk-legs are between three and four times the length of the carapace measured from the base of the rostrum to the base of the great median posterior spine.

The two spines on the sternum between the bases of the second pair of legs may be distinct or indistinct.

The abdomen consists of seven distinct segments in both sexes.

75

In this variable species the constant characters are :—

(1) the discoid (*i.e.*, non-globose) carapace, with elevations only down the middle line:

(2) the long slender legs of the second pair.

(3) the large size of the spine at the external angle of the buccal frame.

In the Museum collection are specimens from the Sandheads, R. Hughli; Mergui; Andamans; and also from Hong Kong, whence the species was originally described by Stimpson.

Doclea muricata (Herbst), Edw.

Cancer muricatus, Herbst, Krabben, I. ii. 211, tab. xiv. fig. 83; and Fabricius, Ent. Syst. II. 459.

Inachus muricatus, Fabricius, Supplement, p. 355.

[*Maia muricata*, Bosc, I. 255.]

Doclea muricata, Milne-Edwards, Hist. Nat. Crust. I. 295.

Doclea muricata, Adams & White, ' Samarang' Crustacea, p. 8.

Doclea muricata, E. Nauck, Zeits. Wiss. Zool., XXXIV. 1880, p. 38, (gastric teeth).

Doclea muricata, C. W. S. Aurivillius, Kongl. Sv. Vet. Akad. Handl., XXIII. 1888-89, No. 4, p. 43, pl. iv. fig. 5.

Doclea muricata, A. O. Walker, Journ. Linn. Soc., Zool., XX. 1890, p. 109.

Doclea muricata, Henderson, Trans. Linn. Soc., Zool. (2) V. 1893, p. 342.

Body and legs, except the hands and dactyli, closely covered with crisp very short velvet.

Carapace subglobular. Rostrum short, distinctly bifid. Besides the spine formed by the basal antennal joint, and two denticles at the outer angle of the buccal frame, the antero-lateral margin is armed with four spines, the last of which, situated near the middle of the branchiostegal border, is considerably the largest. The carapace is traversed fore and aft in the middle line by a row of sharp spines, the last of which, situated on the posterior border, is considerably the largest. Between the median and lateral rows of spines, on the branchial region on either side, are two large spines, one behind the other. There are thus five series of spines upon the carapace, which is otherwise characterized by the distinct delimitation of its regions, and by a sort of festooning of the border between the median and lateral series of regions.

The chelipeds are slender throughout in both sexes, and are hardly equal in length to the carapace measured from the base of the rostrum to the base of the posterior spine: the second pair of trunk-legs are rather more than twice the length of the chelipeds.

The abdomen consists of seven distinct segments in the male; and of four in the female, the 3rd to the 6th being fused.

76

Of 24 specimens from different parts of India there is not one of great size, nor a single adult female.

I believe that this species is only the young form of *Doclea hybrida.*

Doclea hybrida (Fabr.), Edw.

Inachus hybridus, Fabricius, Supplement, p. 355.
[*Maia hybrida,* Bosc, I. 256]; and Latreille, Hist. Nat. Crust., VI. 99.
Doclea hybrida, Milne-Edwards, Hist. Nat. Crust, I. 294.
Doclea hybrida, Adams and White, ' Samarang' Crustacea, p. 7.
Doclea hybrida, Bleeker, Recherches Crust. Ind. Archipel., p. 9.
Doclea hybrida, De Man, Mergui Crust., Journ Linn. Soc., Zool., XXII. 1888, p. 9.
Doclea hybrida, Henderson, Trans. Linn. Soc., Zool. (2) V. 1893, p. 342.
? *Doclea hybridoidea,* Bleeker, Recherches Crust. Ind. Archipel., p. 8.

This species differs from *Doclea muricata*, only in the following characters, which, I think, are merely due to age : —

(1) it is much larger ;

(2) the spine of the antero-lateral series is (except in small females) the smallest, and tubercles are found instead of spines on the dorsal surface of the carapace, the tubercles corresponding in number and position with the spines of *D. muricata* ;

(3) the chelipeds in the adult male are nearly as long as the carapace and rostrum, and have the hands enlarged.

As in *D. muricata* the female abdomen consists of four segments.

As Fabricius, *loc. cit.*, says of this species compared with *D. muricata, vix distinctus videtur.*

We have 29 good specimens from different parts of India, all being large males and egg-laden females. I think that they can only be the adult stage of *Doclea muricata.*

Doclea tetraptera, A. O. Walker.

Doclea tetraptera, A. O. Walker, Journ. Linn. Soc., Zool., Vol. XX. 1890, p. 114 pl. vi. figs. 4–8.

Body and legs, except the hands and dactyli, covered with a dense stiff fur, so stiff on the trunk-legs as to give their joints, though cylindrical, a sharply quadrangular or triangular sectional form.

The circular form of the carapace is a good deal obscured by the unusual development of the rostrum and of the lateral-epibranchial and postero-median spines.

The rostrum is from one-fourth to two-fifths the length of the carapace proper, and ends in two widely divaricated spinules.

In addition to the tooth formed by the basal antennal joint, and to a stout tooth at the angle of the buccal frame, the antero-lateral margin bears four large spines : of these, one, situated on the pterygostomian region, is turned downwards to assist in forming a pterygostomian canal somewhat similar to that of *Doclea canalifera*, etc. : of the other three, which are situated on the branchiostegal region, the last is by far the longest and stoutest—being from one-third to half the length of the carapace—and is directed a little backwards and upwards. Down the middle line of the carapace runs a row of spines, increasing in size from before backwards to the last, which, situated on the posterior border, consists of two branches, one branch directed vertically upwards, the other directed horizontally backwards, the horizontal branch being often half the length of the carapace proper.

On the anterior part of the branchial region, midway between the middle line and the lateral border of the carapace, is a stout spine, visible without any denudation.

The chelipeds in the adult male are equal in length to the carapace and rostrum, and have the hands much broadened, inflated, and very elegantly carinated along the lower border, and the fingers evenly denticulated but not closely apposable in all their extent. In the female the chelipeds are not much more than half as long as the carapace *plus* rostrum and posterior spine, and are rather slenderer than the other legs, the fingers also being closely apposable throughout. In young males, of the size figured by Mr. Walker, the enlargement of the hands is much less marked than in old males.

The second pair of trunk-legs, which are the longest, are from twice to $2\frac{1}{2}$ times the length of the carapace measured from the base of the rostrum to the base of the great postero-median spine.

The sternum in the male has a pair of sharp teeth on its first segment.

The abdomen in both sexes consists of seven separate joints.

Colours in life : dull chocolate, spines white-tipped, chelipeds ivory tinged with pink, legs brownish pink with bright red dactyli.

This species, of which we have a very fine old male, two younger males of different sizes, an adult female, and a young female, appears to be extremely close to *D. calcitrapa*, White (Proc. Zool. Soc., 1847, p. 56 ; Ann. Mag. Nat. Hist., Vol. XX. 1847, p. 61 ; and ' Samarang ' Crustacea, p. 7, pl. i. fig. 2). It appears to differ from *D. calcitrapa* only in the proportions of the legs, which are slender and very long in the last-named species.

It may be mentioned that the rostrum and great spines of the carapace are, judging from the state of two of our specimens, liable to be broken and only very imperfectly repaired again.

7_8

Our specimens all came from the vicinity of the mouth of the River Hooghly.

Alliance II. LISSOIDA.

HOPLOPHRYS, Henderson.

Hoplophrys, Henderson, Trans. Linn. Soc., Zool., Vol. (2) V. 1893, p. 346.

Carapace subovate (elongate pentagonal), with the regions moderately defined and the surface spinose. The rostrum is composed of two short, flattened, acute, divergent spines. The commencing orbits are formed by a supra-ocular eave which has its antero-external angle very strongly and acutely produced, and which is in close contact with a slightly excavated post-ocular tooth, only a very narrow fissure being left between: below, there is no trace of an orbital floor. The eyes are short, and even when fully retracted the cornea is hardly at all concealed from dorsal view. The basal antennal joint is very acutely triangular, the spinous termination being distinctly visible from above: the very short slender mobile portion of the antenna is exposed. The antero-external angle of the merus of the external maxillipeds forms a foliaceous lobe: the merus therefore is broader than the ischium; the palp is attached to its internal angle. The trunk-legs are strongly spinose: the chelipeds, even in the adult male, are slender, but still differ from those of the female in having the fingers more arched and closely apposable only in the distal half.

The abdomen in the male consists of seven distinct segments; but in the female of only five—the fourth to the sixth being fused together.

Hoplophrys oatesii, Henderson.

Hoplophrys oatesii, Henderson, Trans. Linn. Soc. Zool., 1893, p. 347, pl. xxxvi. figs. 1–4.

The gastric region of the carapace is prominent, with two curved rows of spines, the front row (convex anteriorly) consisting of seven spines of which the middle one is the largest, the back row (slightly convex posteriorly) consisting of three spines of which the middle one—the largest of all the spines on the gastric area—is compressed laterally. On the cardiac area, as well as on the gastric area, are two spines placed side by side. On either branchial area are three spines arranged in a triangle, of which the anterior is the largest of all the spines on the carapace, while the most external, which occupies the lateral epibranchial angle, is the most acute and is also unequally bifid. There are also two or three spinules on the hepatic area. Between the

79

spines the surface is perfectly smooth and polished, although there are some tufts of stiff clean hairs.

The rostrum, which consists of two very acute and slightly divergent teeth, is about one-fourth the length of the carapace proper.

The supra-ocular eave is produced forwards as a very acute spine, the base of which is surmounted by a secondary spine. The cornea is surmounted by a spinule.

The chelipeds have the merus slightly, and the carpus strongly spiny, and are equal to the carapace (without the rostrum) in length: they are almost alike in the adults of both sexes, the fingers only of the male differing from those of the female in being closely apposable only in the distal half, instead of throughout. The ambulatory legs, which are about equal to the chelipeds and to one another in length, have the merus carpus and propodite spiny, and the dactylus stout, claw-like, and denticulated on part of the posterior margin.

In the Museum collection are an adult male and an egg-laden female taken by myself, off the Ganjam Coast in 15-25 fms., from a colony of *Spongodes*. The *Spongodes* which belongs to a species (I think new) intermediate in character between *S. cervicornis* and *S. pustulosa*, W. and S., is one of those with a brilliant white cœnosarc and pink zooids, so that the crabs with their porcelain-white bodies, pink spines, and pink-banded legs were with difficulty detected.

Dr. Henderson considers the above species to be closely related to *Schizophrys* and *Microphrys*, but it appears to me to be much more closely related to *Pisa* and *Tylocarcinus*.

TYLOCARCINUS, Miers.

Tylocarcinus, Miers, Journ. Linn. Soc., Zool., Vol. XIV. 1879, p. 664. (*Pisa*, Latr. *part.* ; *Pisa*, Edw. *part.* ; *Milnia*, Stimpson *part.* ; *Microphrys*, Edw. *part.*)

Carapace tuberculated, pyriform, without lateral spines. The rostrum consists of two slender slightly divergent spines.

The eye-stalks are short and are retractile, but not to such an extent as to completely conceal the cornea. The commencing orbits are formed by a supra-orbital cave, the anterior angle of which is produced forwards as a spine roughly parallel with the rostrum, and of a strongly cupped post-ocular process which, instead of being isolated, is in the closest contact above with the supra-ocular cave and below with the basal antennal joint. The basal antennal joint, which is of no great breadth, has its antero-external angle produced to form a sharp tooth, which is not visible from above: the mobile portion of the antenna, which is short, is completely exposed.

The external maxillipeds have the merus as broad as the ischium, and the palp attached to the internal angle of the merus.

The chelipeds in the adult male are somewhat stouter than the other legs, have the palm short and enlarged, and the fingers arched and meeting only at tip : in the female they are slenderer than the other legs, have the palm slender, and the fingers closely apposable throughout. The ambulatory legs are stout, and have the dorsal surface sharply nodose or coarsely spinose.

The abdomen in both sexes consists of seven distinct segments.

This genus, which appears to me to be but slightly distinct from *Pisa* (e.g., *Pisa corallina*), Riss., shows the transition towards *Tiarinia* in the next group.

That it should be grouped with *Tiarinia* and *Macrocoeloma*, as it is by Miers (*loc. cit.*), I cannot agree, since *Tiarinia* has complete orbits and an enormously broad basal antennal joint, which *Tylocarcinus* has not.

The type of *Tylocarcinus*, namely *T. styx* (Herbst) = *Micrphrys styx* A. Milne-Edwards, is placed by the latter author (Nouv. Archiv. du Mus., VIII. 1872, p. 247) between *Picrocerus* and *Criocarcinus* on the one hand and *Hyastenus* on the other ; and this seems to me to be a very natural position.

Tylocarcinus styx (Herbst).

Cancer styx, Herbst, Krabben, III. iii. 53, pl. viii. fig 6 ("nur klein").

[*Pisa styx*, Latr. Encyc., X. 141.]

Pisa styx, Milne-Edwards, Hist. Nat. Crust. I. 308.

Arctopsis styx, Adams and White, 'Samarang' Crust, p. 10 ; and A. Milne-Edwards, in Maillard's L'ile Reunion, Annexe F, p. 6.

Milnia styx, Stimpson, Ann. Lyc. Nat. Hist. New York, Vol. VII. 1862, p. 180.

Micrphrys styx, A. Milne-Edwards in Archiv. du Mus. VIII. 1872, p. 247, pl. xi. fig. 4.

Tylocarcinus styx, Miers, Ann. Mag. Nat. Hist. 1879, Vol. IV. p. 14.

Pisa styx, Richters, Möbius, Meeresf. Maurit., p. 141.

Tylocarcinus styx, de Man, Notes Leyden Mus., Vol. III. 1881, p. 94; and Archiv. fur Naturges. LIII. 1887, p. 228; and Ortmann, Zool. Jahrb. Syst. etc. VII. 1893, p. 62; and Henderson, Trans. Linn. Soc., Zool., 1893, p. 349.

Carapace subpyriform and covered with rounded tubercles, among which the following are distinct :—two in the inter-orbital space ; four in a transverse series on the front part of the gastric region, followed by three in a triangle ; one in the groove between the gastric and cardiac regions, and three in a triangle on the latter region ; two, side by side, on the intestinal region ; and three on the posterior margin. Besides these there are several on either hepatic region, and many on the branchial regions.

The rostrum, which is between one-third and one-fourth the length of the carapace proper, consists of two divergent spines fused together at the base and slightly incurved towards the tip. The anterior angle of the supra-ocular eave is produced forwards as a sharp spine.

The chelipeds in the adult male are equal to the length of the carapace behind the bifurcation of the rostral spines: they are hardly stouter than the other legs, except as to the palm, which is short and inflated : the fingers, which are three-fourths the length of the palm, are strongly arched, and meet only at the tip.

In the female the chelipeds are not quite as long as the post-orbital portion of the carapace, are slenderer than the other legs, and have the palm slender and the fingers closely apposable throughout.

The ambulatory legs are short and stout: the first pair, which are considerably the longest, are rather longer than the carapace and rostrum: the merus and carpus in all are nodose on the dorsal surface, and the dactyli are strong and claw-like: always in the first pair, and sometimes in the succeeding pairs, the merus has a row of coarse spines along its front margin, and the carpus a single stout spine.

Herbst's figure is either a young male, or, more probably, a female. The figure given by A. Milne-Edwards (*loc. cit.*) is very correct; but I do not see how Miers, who cites this figure with affirmation, can call the chelipeds in the male slender: they are, like the other legs, stout, and the hands are distinctly massive.

In the Museum collection are specimens from Ceylon, from the Andamans, and from Mergui; as well as an adult male and female from Samoa obtained from the Museum Godeffroy.

Sub-family IV. MAIINÆ.

Eyes either (1) with orbits, which are either incomplete or complete, but are always complete enough to entirely conceal the cornea, when fully retracted, from dorsal view; or (2) but partially protected by a huge horn-like or antler-like supra-ocular spine, or by a large jagged post-ocular tooth, or by both.

The orbit in the first case is formed in one of two ways : there is always an arched supra-ocular eave, and a prominent post-ocular spine ; and either the interval between the eave and the spine is filled by an intermediate spine which completes the orbital roof; or the supra-ocular eave and the post-ocular process are in close contact with one another, and with a process of the basal antennal joint below, so as to more or less complete the floor also of the orbit.

The basal antennal joint is always very broad, and either has its outer angle produced to aid in forming the floor of the orbit, or is armed distally with one or two large spines.

82

The external maxillipeds have the merus as wide as or much wider than the ischium, and the palp inserted at the antero-internal angle of the merus.

The rostrum is formed of two spines, which may be horizontal, semi-deflexed, or completely deflexed; in the last case the spines are usually more or less fused together.

The ambulatory legs are of no great length.

Key to the Indian genera.

Alliance 1. MAIOID-A.—C a r a p a c e either regularly pyriform or sub-circular: rostral spines horizontal: orbits incomplete below; but fairly well roofed in above (1) by a su-pra-ocular eave, which has at least its postero-exter-nal angle pro-duced, (2) by a post-ocular spine, and (3) by a spine intercalated between (1) and (2).	1. S u p r a - o c u l a r eave and interme-diate spine very prominent: eye-stalks slender and curved, with the cornea elongate and occupying a position more ven-tral than terminal.	i. The antennulary flagellum springs, or appears to spring, from with-in the orbit......... MAIA.
		ii. The antennulary flagellum arises quite clear of the orbit................ PARAMITHRAX. [CHLORINOIDES.]
	2. S u p r a - o c u l a r eave and interme-d i a t e spine dis-tinct, but not very prominent: eye-stalks stout, with rounded cornea which occupy a position as much terminal as ven-tral.	i. Carapace p y r i-form: r o s t r a l spines of consider-able length, and with one or more accessory spines on the outer sur-face............... SCHIZOPHRYS.
		ii. Carapace subcir-cular: r o s t r a l spines simple, and so short as to hardly break the general outline of the carapace....... CYCLAX.

Alliance 2. STENOCIONOPOIDA.—Carapace pyriform, often broadened anteriorly: the orbits either have the form of long semitubular antlers which sheathe the eye-stalk, but do not protect the eye, the cornea in retraction being protected by the base of an extremely long and promi-nent, isolated, post-ocular horn; or are reduced to the form of long outstanding horns similar to those of the rostrum: eye-stalks extreme-ly long: the external maxilli-peds have the external angle much produced: the rostrum consists of two long horns.	1. Orbits in the form of huge semi-tubular antlers followed by a long isolated post-ocular tooth: rostrum vertically deflexed: buccal frame much broader in front than behind.	CRIOCARCINUS.
	2. Orbits in the form of long outstanding horns similar to those of the rostrum, which is not deflexed, buccal frame quadrangular..................	STENOCIONOPS.

Alliance 3. PERICER-	1.	Carapace oblong: rostrum broadly laminar, vertically or nearly vertically deflexed : orbits complete, but shallow..		MICIPPA.
OIDA.— Carapace usually broadened anteriorly by the outstanding or-				
bits : the o r b i t s are either nearly or quite complete above and below,	2.	Carapace subcylindrical, the rostrum *along with the front part of the gastric region* vertically deflexed......		CYPHOCARCINUS.
being formed by a strongly-arched supra-ocular eave in close contact with an excavated	3.	Carapace more or less pyriform: rostral spines distinct from the base, horizontal or slightly	i. Rostral spines divergent..	MACROCOELOMA.
post-ocular lobe, a process of the basal a n t'e n n a l joint filling in the floor below.		deflexed: orbits in the form of outstanding [t u b e s which, completely ensheathe the eyes.	ii. Rostral spines parallel and closely approximated throughout their extent..........	TIARINIA.

Alliance I. MAIOIDA.

MAIA (Lamk.) Edw.

[*Maia*, Lamarck, Syst. Anim. sans verteb. V. 154 (*partim*).]
Maia, Latreille, Hist. Nat. Crust. VI. 87 (*partim*).
Maia, Desmarest, Consid. Gen. Crust., p. 143.
Maia, Milne-Edwards, Hist. Nat. Crust., I. 325.
Maia, Miers, Journ. Linn. Soc., Zool., Vol. XIV. 1879, p. 655.

Carapace pyriform, with the regions indistinct, the surface closely granular or spinular, and the lateral borders usually armed with large spines. The rostrum consists of two rather short, straight, divergent spines. The basal joint of the antennæ is broad, and has both the antero-external and antero-internal angle produced to form spines : the mobile portion of the antenna, which appears to spring from within the orbit, is completely exposed. The eye-stalks are long and curved, and bear the cornea chiefly on their ventral surface. The orbit is formed by a prominent supra-ocular eave which has its postero-external angle produced, by a sharp post-ocular spine, and by another spine between these two : the eyes are completely concealed from dorsal view when retracted. The external maxillipeds have the merus as broad as the ischium, the palp being attached to the antero-internal angle of the merus.

The chelipeds are slender, with cylindrical joints and styliform fingers. The ambulatory legs decrease very gradually in length : the first pair are not much longer than the carapace and rostrum : the dactyli of all are styliform.

The abdomen in both sexes consists of seven distinct segments.

84

Maia spinigera, de H.

Maia spinigera, de Haan, Faun. Japon. Crust., p. 93, pl. xxiv. fig. 4.
Maia spinigera, Adams and White, ' Samarang' Crustacea, p. 15.
Maia spinigera, Dana, U. S. Expl. Exped. Crust., pt I. p. 85.
Maia spinigera, Ortmann, Zool. Jahrb. Syst. &c., VII. 1893, p. 51.

Carapace armed with long spines along the antero-lateral borders, down the median line, and in an oblique series on either branchial region joining the median to the antero-lateral series. Excluding the pre-ocular and post-ocular spines and the spines between them, there are four large spines on the antero-lateral border : and there are three large spines in an oblique series on either branchial region. In the middle line of the carapace there are in the gastric region two spines, in the anterior cardiac one, in the post-cardiac one, in the intestinal one, and on the posterior border a pair. Between these large spines the surface of the carapace is sharply, finely, and evenly granular.

The rostrum consists of two moderately divergent spines, the length of which is about one-fourth that of the carapace.

The chelipeds are smooth and very slender, and are rather shorter than the 2nd pair of trunk-legs : the latter, which are the longest of all, are about one-sixth longer than the carapace and rostrum. The merus of all the ambulatory legs has a strong spine at the distal end of its upper border : all the joints of all the ambulatory legs are covered with long hairs.

In the Museum collection is a single specimen from the coast of Beluchistán.

Maia gibba, n. sp. Plate IV. fig. 5.

Very near *Maia miersii*, Walker (J. L. S., Zool., Vol. XX. 1890, p. 113, pl. vi. figs. 1–3.

Distinguished (1) by the globose inflation of the posterior (branchiostegal) part of the closely and crisply tubercular carapace, and by the corresponding declivity of the anterior part, giving the animal a hunch-backed appearance ; (2) by the absence of *large* marginal spines on the carapace.

Carapace remarkably swollen in its posterior part, where its greatest breadth is from about three-fourths (♂) to seven-eighths (♀) its extreme length with the rostrum; and closely covered with sharp piliferous tubercles, which, in the male, but hardly in the female, become spinular in the middle line and along the lateral borders.

The rostrum, which, like the anterior part of the carapace, is somewhat declivous, ends in two acute divergent hairy spines, which in the

85

male are about one-sixth, in the female about one-eighth, the rest of the carapace in length. The eyes and orbits are just as in *M. squinado* (with specimens of which this species has been compared), only the cornea is relatively very much larger, and almost entirely ventral, in the present species, and the spine between the spine of the pre-orbital-hood and the post-orbital spine is nearly as large as either of these.

The antennæ are in all respects as in *M. squinado*, except that the basal joint is slightly narrower.

The appendages are just as in *M. squinado*—the legs being short and hairy and the chelipeds smooth and polished—with the single difference that the chelipeds are only as long as, and are much slender-er than the *fifth* pair of legs, and are therefore very much shorter than the second pair, which hardly exceed the carapace and rostrum in length.

	Male.	Female.
Length of carapace	32 millim.	41 millim.
Greatest breadth of carapace	25 ,,	35 ,,
Length of chelipeds	24 ,,	31 ,,
,, ,, 2nd pair of trunk-limbs	33·5 ,,	46 ,,

Loc. Andaman Sea, 250 fms.

PARAMITHRAX, Edw.

Paramithrax, Milne-Edwards, Hist. Nat. Crust. I. 323.
Paramithrax (*Paramithrax* et *Leptomithrax*), Miers, Journ. Linn. Soc. Zool., Vol. XIV. 1879, pp. 655 and 656.
Acanthophrys (partim), A. Milne-Edwards, Ann. Soc. Ent. Fr. (4) V. 1865. p. 140.
Chlorinoides, Haswell *infra;* and Miers *infra.*

Sub-genus CHLORINOIDES, Haswell.

Chlorinoides, Haswell, P. L. S., N. S. Wales, Vol. IV. 1879, p. 442; and Ann. Mag. Nat. Hist., Vol V. 1880, p. 146; and Cat. Austral. Crust., p. 17.
Chlorinoides, Miers, 'Challenger' Brachyura, p. 51.

Carapace pyriform, convex, with the regions indistinct; armed with some very large acute spines. The rostrum consists of two long slender divergent horns. The basal antennal joint is just as in *Maia*, but the mobile portion of the antenna has no connexion with the orbit. The eyes and orbits are as in *Maia*, but the supra-ocular hood has its anterior angle as well as its posterior angle produced into a spine. The external maxillipeds are as in *Maia*, as are also the ambulatory

8G

legs. The chelipeds however differ, at any rate in the male, in which sex they are stouter than any of the other legs, have the palms enlarged, and the fingers arched and meeting only at the tips, which are not excavated.

The abdomen in both sexes consists of seven distinct segments.

As Miers has pointed out ('Challenger' Brachyura, p. 52), *Chlorinoides* may be regarded as a sub-genus of *Paramithrax*, and is also closely connected with *Acanthophrys aculeatus* A. Milne-Edwards (Ann. Soc. Ent. Franc. (4) V. 1865, p. 140, pl. iv. fig. 4). According to Miers, with whom I entirely agree, if *Acanthophrys aculeatus* is the type of the genus *Acanthophrys*, then *Chlorinoides* is synonymous with *Acanthophrys*.

Paramithrax (Chlorinoides) aculeatus, (Edw).

Chorinus aculeata, Milne-Edwards, Hist. Nat. Crust. I. 316.
Chorinus aculeatus, Adams and White 'Samarang,' Crust., p. 13.
Paramithrax (Chlorinoides) aculeatus, var. *armatus*, Miers, Zool. H. M. S. 'Alert,' pp. 182 & 193, pl. xviii. fig. A.
Chlorinoides aculeatus, Miers, 'Challenger' Brachyura, p. 53.
Chorinus aculeatus, C. W. S. Aurivillius, Kongl. Sv. Vet. Akad. Handl., Bd. XXIII. No. 4, p. 38, pl. ii. fig. 7.
Chlorinoides aculeatus, Henderson, Trans. Linn. Soc., Zool., 1893, p. 345.

Carapace pyriform, convex, smooth, armed with five huge thorn-like spines down the middle line, and with two even larger spines on the branchial region : there are also, on either pterygostomian region, two oblique crests, the anterior with three or four teeth—two of which are visible in a dorsal view — the posterior with one or two.

The rostrum consists of two large divergent horns, the length of which is considerably more than half that of the carapace proper.

The orbit consists of a supra-ocular hood, the angles of which (especially the anterior) are strongly produced, of a bilobed post-ocular tooth, and of a long spine filling the interval between the two, just as in *Maia spinigera*. The basal antennal joint, as in most of the forms included in this group, has a strong spine at its antero-external, and another at its antero-internal angle.

The chelipeds in the female are slender, and are only equal to the post-rostral portion of the carapace in length : as in the male, the merus has its crest-like upper and lower edges sharply scallopped and the carpus is cristate above. In the male the chelipeds are stouter than the other legs, especially as to the palm, which is considerably enlarged. The ambulatory legs decrease gradually in length from the 1st pair, which are equal in length to the carapace *plus* two-thirds of the rostrum : the merus in the first two pairs has a very strong spine at the

87

distal end of its upper border ; but this in the case of the last two pairs
is often reduced to a tubercle.

The body and legs in this species are somewhat hairy and are more
or less encrusted with sponges, zoophytes, polyzoa, etc.

In the Museum collection are specimens from the Arakan Coast,
Mergui, and Ceylon.

Paramithrax (Chlorinoides) longispinus (de Haan).

Maja (Chorinus) longispina, de Haan, Faun. Japon., Crust., p. 94, pl. xxiii. fig. 2.

Chorinus longispina, Adams and White, ' Samarang' Crust., p. 12.

Paramithrax (Chlorinoides) longispinus, Miers, Zoology H. M. S. ' Alert,' pp. 517
and 522.

Chlorinoides longispinus, Miers, ' Challenger' Brachyura, p. 53.

Chlorinoides longispinus, A. Ortmann, Zool. Jahrb. Syst., etc., VII. 1893, p. 53.

This species differs from *P. aculeatus* in the following constant
characters :—

> (1) it is a much smaller species ;
>
> (2) all the spines, including the rostral spines, are elegantly
> knobbed at tip ;
>
> (3) in the median line of spines the third—the one on the cardiac
> region—is cleft transversely into two from the base ;
>
> (4) the two oblique dentate ridges on the pterygostomian region
> are present, but the outermost tooth on the front ridge is
> produced to form a long spine ;
>
> (5) the spine at the anterior angle of the supra-ocular hood is
> similar in size, form, and direction to the other large spines
> of the carapace ;
>
> (6) the rostral spines are less than half the length of the cara-
> pace ;
>
> (7) the antero-external angle of the basal antennal joint is pro-
> duced to form, not a spine, but an elegantly curved folia-
> ceous lobe ;
>
> (8) the meropodites of all the ambulatory legs have the terminal
> spine distinct and knobbed at the tip.

This species commonly encrusts itself with a very regular plate-
armour of Orbitolites and rounded fragments of Nullipore, etc.

In the Museum collection are good series from off Ceylon 33-34
fathoms, from the Andaman Sea down to 41 fathoms, and from the
Madras Coast.

SCHIZOPHRYS, White.

Schizophrys, White, Ann. Mag. Nat. Hist., Vol. II. 1848, p. 282.
Schizophrys, Miers, Journ. Linn. Soc., Zool., Vol. XIV. 1879, p. 660 (*et synon.*);
and 'Challenger' Brachyura, p. 66.
Dione, de Haan, Faun. Japon. Crust., p. 82.

Carapace broadly pyriform, with the surface granular and the lateral margins strongly spinate. The rostrum consists of two short stout slightly incurved spines, the outer border of which carries one or two accessory spines. The orbit is formed by a little-prominent supra-ocular eave, and a sharply bilobed post-ocular tooth, with a broad spine in the interval between the two : the eye-stalks are stout and the cornea terminal, not ventral, in position. The basal antennal joint is somewhat narrowed anteriorly, and ends in two sharp spines—as in the genera immediately preceding : the mobile portion of the antenna is freely exposed. In the external maxillipeds the merus is rather broader than the ischium, and the palp is attached to the antero-internal angle of the merus.

The chelipeds have the merus and carpus granular or spiny ; the palm long, smooth and slender ; and the fingers longitudinally channelled in their distal half—this being specially marked in the adult male, in which also the chelipeds are longer and stouter than the other legs.

The ambulatory legs are stout, have cylindrical joints, and decrease gradually in length.

The abdomen in both sexes consists of seven distinct segments.

Schizophrys aspera, (Edw.)

Mithrax asper, Milne-Edwards, Hist. Nat. Crust., I. 320; and Dana, U. S.
Expl. Exp. Crust., pt. I. p. 97, pl. ii. figs. 4 *a–b*.
Schizophrys aspera, A. Milne-Edwards, Nouv. Archiv. du Mus. VIII. 1872, p. 231,
pl. x. fig. 1 ; and Haswell, Proc. Linn. Soc., N. S. Wales, Vol. IV. 1879, p. 447 ; and
Cat. Austr. Crust., p. 22 ; and Miers, Zool. H.M.S. 'Alert,' pp. 182 and 197,
and 'Challenger' Brachyura, p. 67 ; and De Man, Archiv. fur Naturgesch., LIII.
1887, p. 226, and Journ. Linn. Soc., Zool., Vol. XXII. 1888, p. 20 ; and C. W. S.
Aurivillius, Kongl. Sv. Vet. Akad., Handl. XXIII. 1888-89, No. 4, p. 51 ; [and
Cano, Boll. Soc. Nat., Napol , III. 1889, p. 179]; and A. O. Walker, Journ. Linn. Soc.,
Zool., Vol. XX. 1890, pp. 109 and 113 ; and Ortmann, Zool. Jahrb. Syst , etc., VII.
1893, p. 57; and J. R. Henderson, Trans. Linn. Soc., Zool., (2) V. 1893, p. 346;
and Mary J. Rathbun, Proc. U. S. Nat. Mus., Vol. XVI. 1893, p. 91.
Schizophrys serratus, White, P. Z. S., 1847, p. 223, fig. ; and Ann. Mag. Nat. Hist.,
Vol II. 1848, p. 283, fig. ; and Adams and White, ' Samarang ' Crust., p. 16.
Schizophrys spiniger, White, *ll. cit.;* and Adams and White *loc. cit. ;* and
? Kossmann, Reise Roth. Meer., Crust., p. 15.
Maja (Dione) affinis, de Haan Faun. Japon. Crust., p. 94, pl. xxii. fig. 4; and
Adams and White, ' Samarang ' Crust., p. 15 ; and Stimpson, Proc. Ac. Nat. Sci.,
Philad., 1857, p. 218.

89

Mithrax spinifrons, A. Milne-Edwards, Ann. Soc. Ent., France, (4) VII. 1867, p. 263.

Mithrax affinis, F. de B. Capello, Jorn. Sci., Lisb., 1870-71, p. 264, pl. iii. figs. 4, 4a.

Mithrax (Schizophrys) affinis, triangularis (et varr. *excipe var.* dichotoma) Kossmann, Reise Roth. Meer., Crust., pp. 11 and 13; and *Schizophrys triangularis* var. *indica*, Richters, Möbius, Meeresf. Maurit., p. 143, pl. xv. figs. 8–14.

Carapace pyriform, its greatest breadth about $\frac{9}{10}$ its length behind the point of bifurcation of the rostral spines, its surface closely and unevenly granular, with scattered sharp tubercles in addition. Exclusive of the large unequally-bifid post-ocular spine, the antero-lateral border is armed with six equidistant spines, the last of which is the smallest and is situated on a rather higher level than the others: the posterior border proper is generally beaded, and has its angles produced and upturned.

The rostrum consists of two stout parallel or incurved spines, the length of which is from one-fifth to one-sixth that of the carapace proper, and the outer border of each of which carries a strong accessory spine.

The basal antennal joint ends in two stout spines, and there is a spine on the sub-hepatic region outside the angle of the buccal frame, and a sharp denticle in the middle of the inferior border of the orbit.

The chelipeds vary: in both sexes the palm is long — twice the length of the fingers — smooth, polished, and either quite unarmed, or armed, at the near end of the upper border, with a spine or with two or three denticles; and in both sexes the merus and carpus are either spiny or granular.

But whereas in old males the chelipeds are stouter than any of other legs, are more than half again as long as the carapace and rostrum and nearly half again as long as the 2nd pair of legs, and have deeply channelled fingers that meet in less than their distal half; in females and young males they are not stouter than the other legs, are not quite equal in length to the carapace and rostrum or to the second pair of legs, and have the fingers less deeply channelled, and apposable in at least half their extent.

The ambulatory legs decrease very gradually in length: they have short claw-like dactyli, and the merus is armed at the far end of the upper border with a spine or tubercle. The body and legs are hairy, and the animal frequently protects itself with flat pieces of Nullipore, &c.

In the collection is a large series of specimens from all parts of the Indian coast, from Mergui and Tavoy on the East to Karáchi on the West.

Schizophrys dama, (Herbst.)

Cancer dama, Herbst, Krabben, III. iv. p. 5, tab. lix. fig. 5.
Mithrax dama, Milne-Edwards, Hist. Nat. Crust., I. 319.
Mithrax (Schizophrys) dama, Kossmann, Reise Roth. Meer., Crust., pp. 11 and 13.

This species differs constantly from *Schizophrys aspera* in the following particulars :—

(1) the carapace is much more elongate, its greatest breadth being only about $\frac{3}{4}$ its length behind the point of bifurcation of the rostral spines ;

(2) the rostrum is rather longer, and has *two* accessory spines on its outer border ;

(3) there is no (ventral) spine on the sub-hepatic region ;

(4) the surface of the carapace is more closely and evenly, but more bluntly, granular.

The specimens in the Museum collection come from the Straits of Malacca.

CYCLAX, Dana.

Cyclax, Dana, U. S. Expl. Exp., Crust., pt. I. p. 99.
Cyclomaia, Stimpson, Amer. Journ. Sci. and Arts, Vol. XXIX. 1860, p. 133 ; and A. Milne-Edwards, Nouv. Archiv. du Mus., VIII. 1872, p. 235 (et synon.)
Cyclax (Cyclax and Cyclomaia), Miers, Journ. Linn. Soc., Zool., Vol. XIV. 1879, p. 660.

This genus differs from *Schizophrys,* from which, perhaps, it ought not to be separated, only in the form of the carapace, and in the degradation and shortening of the rostrum, with which is correlated a shortening and broadening of the basal antennal joint. (In one species the legs are slender). The carapace is subcircular ; the rostrum obsolescent and bifid ; the basal antennal joint very short and broad, and armed with a third spine — a very small one, situated on the outer margin.

Cyclax (Cyclomaia) suborbicularis, (Stimpson).

Mithrax suborbicularis, Stimpson, Proc. Ac. Nat. Sci., Philad., 1857, p. 218.
Cyclax spinicinctus, Heller, Crust. Roth. Meer, in SB. Ak., Wien, XLIII. i. 1861, p. 304, tab. i. figs. 7-8 : and Richters, in Möbius, Meeresfauna Maurit., p 144.
Cyclomaia margaritata, A. Milne-Edwards, Nouv. Archiv. du Mus., VIII. 1872, p. 236, pl. x. figs. 2-3 ; and Haswell, P. L. S., N. S. Wales, Vol. IV. 1879, [p. 441, and Cat. Austral. Crust., p. 21.
Cyclomaia suborbicularis, Ortmann. Zool. Jahrb., Syst., etc., VII. 1893, p. 58.
[Cyclomaia margaritata, F. Muller, Verh. Ges., Basel, VIII. p. 473.]

Carapace subcircular, its surface closely beaded, with some larger spinules regularly interspersed : the lateral margin is armed with six

91

large spines (exclusive of the large curved unequally-bifid post-ocular spine) the first of which is often bifid : close to the posterior margin, in the middle line, is a pair of smaller spines.

The rostrum consists of two triangular teeth, which although broader are not longer than the spines of the lateral margin.

The eyes are of moderate length and are retractile into orbits formed, as in *Schizophrys*, *Maia*, etc., of a supra-ocular eave, a large post-ocular spine, with another spine in the interval between the two : the supra-ocular eave has its angles slightly produced and spiniform.

The broad short basal antennal joint ends in two stout teeth, and has a third denticle on its outer margin.

The chelipeds in the female and young male are slightly more slender than the other legs, and are as long as the carapace or as the 2nd pair of trunk-legs *minus* the dactylus : they have a long slender smooth palm, nearly twice the length of the fingers. The ambulatory legs are hairy, have short claw-like dactyli, and decrease gradually in length.

In the Museum collection are specimens from the Madras coast and from the Andamans.

Alliance II. STENOCIONOPOIDA.

CRIOCARCINUS, Edw.

Criocarcinus, Milne-Edwards, Hist. Nat. Crust., I. 331.
Criocarcinus, Miers, Journ. Linn. Soc., Zool., Vol. XVI. 1879, p. 661.

Carapace shaped and armed much as in *Chlorinoides*, but with the hepatic regions concave as in *Micippe*. The rostrum consists of two curved almost vertically deflexed spines, which are fused together in their basal half. The eye-stalks are slender and of extreme length. The orbit is formed of a semi-tubular branching supra-ocular hood which encloses the eye-stalk, and of a long slender post-ocular spine, against the base of which the eye is retractile : the supra-ocular hoods have the appearance of a pair of antlers. The basal antennal joint is broad, and has a strong spine at either anterior angle : the mobile portion of the antenna is freely exposed.

The buccal frame is narrow behind and broad in front, as in *Micippe*; and the merus of the external maxillipeds is broader than the ischium, and carries the palp at its deeply-notched internal angle.

The chelipeds are shorter, and in the male somewhat stouter but in the female somewhat slenderer, than the other trunk-legs, which again are of no great length and decrease gradually from the 2nd pair.

The abdomen consists of seven distinct segments in the male, of five in the female.

Criocarcinus superciliosus (Herbst), Guérin, Edw.

Seba, III. xviii. 11 : Linnæus, Syst. Nat., I. 2, 1047, No. 45.
Cancer superciliosus, Herbst, Krabben, I. ii. 227, tab. xiv. fig. 89.
Criocarcinus superciliosus, Guérin, Voy. Coquille, Zool., Vol. II. Crust., p. 19.
Criocarcinus superciliosus, Milne-Edwards, Hist. Nat. Crust., I. 332.
Criocarcinus superciliosus, A. Milne-Edwards, Nouv. Archiv. du Mus., VIII. 1872,
p. 242, pl. xi'. fig. 3.
Criocarcinus superciliosus, Kossmann, Reise Roth. Meer., Crust., p. 10, tab. iii.
fig. 6 (*vide synon*).

Carapace pyriform, broadened anteriorly by the antler-like "orbits,"
with the hepatic regions sunken, and the other regions fairly distinct:
in addition to numerous pearly tubercles, which are tufted with curly
bristles, the carapace is armed with several large knob-tipped spines,
namely two in the middle line on the gastric region, one in the middle
line on the posterior border, one on either side near the boundary of
the hepatic and branchial regions, and one, directed obliquely backwards, near the middle of either branchial region.

The rostrum consists of two vertically deflexed spines, the bases of
which are broadened and fused together, and the points of which are
divergent and elegantly curved.

The eyes and orbits have already been described in a general way:
the long semi-tubular supra-ocular hood ends in three diverging tines,
and the long post-ocular spine has its anterior border armed with two
or three denticles.

The external maxillipeds have the outer edge thin and sharp, the
outer edge of the ischium being emarginate, and the outer angle of the
merus being produced.

The chelipeds are shorter than the other trunk-legs, and are about
as long as the carapace behind the level of the post-ocular spine. In
the male they are slightly stouter than the other legs, and have the
palm a little swollen : in the female they are slenderer than the other
legs, and have the palm slender and a little tapering.

Of the ambulatory legs, which are hairy, the first two pairs are
slightly the longest, both being rather less than one-third longer than
the post-rostral portion of the carapace: the last two pairs are not
much shorter.

In the Museum collection are specimens from the Andaman Islands.

STENOCIONOPS, Latr.

[*Stenocionops*, Latreille, R. A., (2) IV. 59.]
Stenocionops, Milne-Edwards, Hist. Nat. Crust., I. 337.

" Carapace narrow, uneven, and armed posteriorly with a large
triangular prolongation which covers the base of the abdomen. The
93

rostrum is formed of two styliform divergent horns. The supra-ocular border is armed with a horn similar to those of the rostrum, but directed more obliquely. The eye-stalks are slender, immobile and extremely salient; their length is half the greatest breadth of the body. The first joint of the antennæ is much longer than broad, the second is slender and is inserted beneath the rostrum.

The epistome is nearly square, and the external maxillipeds have the merus extremely dilated at the antero-external angle, and excavated at the antero-internal angle. The trunk-legs, in the female, are slender and cylindrical: those of the first pair (chelipeds) are hardly stouter and are much shorter than the second, which latter are a little longer than the carapace and rostrum : the others diminish very gradually in length : all the ambulatory legs have sharp, recurved dactyli. The abdomen of the female consists of five segments, the 4th, 5th and 6th segments being fused together." (Edw.)

Stenocionops cervicornis (Herbst).

Cancer cervicornis, Herbst, Krabben, III. iii. 49, pl. lviii. fig. 2.
[*Stenocionops cervicornis,* Guérin, Icon. Regne An., Crust., pl. 8 bis, fig. 3].
Stenocionops cervicornis, Milne-Edwards, Hist. Nat. Crust., I. 338.
Stenocionops cervicornis, Cuvier, Regne Animal, Crust., pl. xxxi. fig. 1.
Stenocionops cervicornis, and ? *curvirostris,* A. Milne-Edwards, Ann. Soc. Ent., France, (4) V. 1865, p. 135 (pl. v. figs. 1-1e.)
Stenocionops cervicornis, E. Martens, Verh. zool. bot. Ges., Wien, XVI. 1866, p. 379.
[*Stenocionops cervicornis,* Cano, Boll. Soc. Nat., Napol., III. 1889, p. 177.]
Stenocionops cervicornis, Henderson, Trans. Linn. Soc., Zool., 1893, p. 343.

" Carapace uneven and tuberculated : rostral and supra-ocular horns slender, very long, and nearly co-equal : two large conical elevations on the sides of either hepatic region : antennæ shorter than the rostrum : chelæ finely toothed and a little incurved : legs smooth." (Edw.)

Alliance III. PERICEROIDA.

MICIPPA, Leach.

Micippa, Leach, Zool. Miscell., III. p. 16.
Micippe, Desmarest, Consid. Gen. Crust., p. 148.
Micippe, Milne-Edwards, Hist. Nat. Crust., I. 329.
Micippa, Miers, Journ. Linn. Soc., Zool., Vol. XIV. 1879, p. 661 ; Ann. Mag. Nat. Hist., Vol. XV. 1885, p. 3 ; and ' Challenger ' Brachyura, p. 69.

Carapace nearly oblong, depressed, rounded behind, broadened anteriorly, and ending at a broad, lamellar, more or less vertically deflexed rostrum, the tip of which is cleft or emarginate. The eye-stalks are long, and the corneæ, which are rather ventral than terminal in position, can be completely retracted from dorsal and usually also from ventral view. The orbit is formed by a sharply-arched supra-ocular eave, which is in contact either with an excavated post-ocular spine or with an intercalated spine as in *Maia*, and is partly or entirely completed below and in front by a process of the broad basal antennal joint. The mobile portion of the antenna is completely exposed.

The buccal frame is broadened in front : the merus of the external maxillipeds is broader than the ischium, and has its external angle expanded and its internal angle notched for the insertion of the palp.

The chelipeds in the adult male are as long as or a little longer than the carapace, are a little stouter than the other legs, and have the palm broader than the other joints, and the fingers arched to meet only at the tip. The chelipeds in the female are slenderer than the other legs, are about the same length as the carapace, and have slender palms and almost straight fingers. The ambulatory legs are moderately elongate, subcylindrical, and have the dactyli not much or not at all shorter than the propodites.

Abdomen, in both sexes, seven-jointed.

Key to the Indian species of Micippa.

I. Rostrum very broad, ending in four sharp lobes or spines
    (*i.e* , each lobe of the rostrum bilobed)......................... *M. philyra.*

II. Rostrum moderately broad, ending in two long sharp lobes
    or spines (*i.e.*, each lobe of the rostrum simple), not
    inflexed at tip.............. .... *M. thalia.*

III. Rostrum moderately broad, inflexed at tip; ending in two
     insignificant blunt lobes, each of which has a small
     tooth at its external angle :—

    1. Three large pearl-like tubercles embedded
        in the posterior margin....................... *M. margaritifera.*

    2. Two small pearl-like tubercles embedded
        in the posterior margin, with a group of
        small spinules between them.............. *M. margaritifera*
                          var. *parca.*

Micippa philyra, (Herbst.) Leach.

Cancer philyra, Herbst, Krabben, III. iii. p. 51, pl. lviii. fig. 4.
Micippa philyra, Leach, Zool. Miscell., III. 16; and Desmarest, Consid. Gen. Crust., p. 149, pl. xxii. fig. 2 ; and Guérin, Icon. R. A., pl. viii bis, fig. 1 ; and Milne-Edwards, Hist. Nat. Crust., I. 330 ; and Adams and White, ' Samarang ' Crust., p. 15 ; and A. Milne-Edwards, Nouv. Archiv. du Mus., VIII. 1872, p. 239, pl. xi. fig. 2 and Kossmann, Reise Roth. Meer., Crust., p. 6 (*ubi synon.*) ; and varr. *platipes* and

95

mascarenica, pl. iii. figs. 2-3; and Richters, Möbius, Meeresfauna, Mauritius, p. 143,
pl. xv. figs. 6-7, and var. *latifrons*, p. 142, pl. xv. figs. 1-5; and Lenz and Richters,
Abh. senck. Ges. XII. 1881, p. 421; and Miers, Zoology H. M. S. ' Alert,' pp. 182
and 198, and Ann. Mag. Nat. Hist., 1885, Vol. XV. p. 6, and 'Challenger' Brachyura,
p. 69; and Ortmann, Zool. Jahrb. Syst., &c., VII. 1893, p. 59; and J. R. Henderson,
Trans. Linn. Soc., Zool., 1893, p. 348.

 Micippe platipes, Rüppell, Beschrib. und Abbild., 24 Krabben Roth. Meer.,
Frankfort, 1830, p. 8, tab. i. fig. 4; and Milne-Edwards, Hist. Nat. Crust., I. 333
(*Paramicippe*); and Heller, Crust. Roth. Meer., SB. Ak., Wien, XLIII. 1861,
p. 299, tab. i. fig. 2; and De Man, Archiv. fur Naturgesch., LIII. 1887, p. 227
(*Paramicippe*).

 Micippe bicarinata, Adams and White, ' Samarang ' Crust., p. 16, (*sec.* Kossmann
and Miers).

 ? *Micippe hirtipes*, Dana, U. S. Expl. Exp., Crust., pt. I. p. 90, pl. i. figs. 4 *a–e*;
and Stimpson, Proc. Ac. Nat. Sci., Philad., 1857, p. 218; and Heller, Reise ' Novara,'
Crust., p. 3.

 Micippa spatulifrons, A. Milne-Edwards, Nouv. Archiv. du Mus., VIII. 1872,
p. 240, pl. xi. fig. 3; and Haswell, Proc. Linn. Soc., N. S. Wales, Vol. IV. 1879,
p. 445, and Cat. Austral. Crust., p. 24.

 Micippa mascarenica, Kossm., Miers, Ann. Mag. Nat. Hist., 1885, Vol. XV.
p. 7, and 'Challenger' Brachyura, p. 69; and A. O. Walker, Journ. Linn Soc., Zool.,
Vol. XX. 1890, p. 109; and J. R. Henderson, Trans. Linn. Soc., Zool., 1893, p. 348.

 Micippa superciliosa, Haswell, Proc. Linn. Soc., N. S. Wales, Vol. IV. 1879,
p. 446, pl. xxvi. fig. 2, and Cat. Austral. Crust., p. 25.

 Paramicippa asperimanus, Miers, Zoology H. M. S. ' Alert,' pp. 517 and 525.

  Body and ambulatory legs closely covered by a woolly tomentum.
Carapace with the regions well defined by smooth sulci, the hepatic
regions sunken and pinched in, the surface closely and unevenly
granular : the lateral margins are armed with knob-tipped spinules,
of which there are sometimes as many as six, sometimes as few as two,
on either side.

  The rostrum consists of a broad lamina which in the female is
quite vertically, but in males is not so much deflexed, its sides are
gently sinuous, and it ends in four sharp-cut lobes. The eyes are
completely retractile within the orbit.

  The basal antennal joint is short and is extremely broad anteriorly,
its greatly produced antero-external angle completing the orbit below
and in front. The mobile portion of the antenna, which is freely
exposed, varies in length and in the form of the flattened 2nd joint
of the peduncle. In some males (var. *mascarenica*) the mobile portion
of the antenna is half the length of the horizontal portion of the
carapace, and the length of the 2nd joint is rather more than one-third
the breadth of the rostrum at its own point of origin. But in all
ovigerous females, and in certain males, the mobile portion of the
antenna is between one-third and one-fourth the length of the hori-

rontal portion of the carapace, and the length of the 2nd joint is less than one-third the breadth of the rostrum at its own point of origin — the joint also being somewhat broadened.

The chelipeds also vary. In certain males, both adult and young (var. *mascarenica* partim), they are stouter than the other legs, are very variably granular, are a little longer than the carapace, have the hand very variably broadened and inflated, and the fingers closely apposable only at tip.   In all females they are a little shorter than the carapace, are quite smooth, are rather slenderer than the other legs, and have slender palms, and fingers that are closely apposable in the greater part of their extent.   In certain other adult males they are intermediate in condition, approaching more to the female type.

The ambulatory legs are moderately stout and are hairy : the 1st pair, which are the longest, are rather longer than the chelipeds ; the others decrease gradually in length.

Miers' valuable paper, Ann. Mag. Nat. Hist., 1885, Vol. XV. pp. 6–8 should be consulted. After examining over forty specimens from the Andamans I adhere to Kossmann's synonomy and opinion (*loc. cit.*)

The characters upon which the separation of *M. mascarenica* from *M. philyra* is based are all variable; and I think that we have here to deal with a case of male dimorphism, such as is known to occur in certain Beetles, where one form of male is aberrant from the female type while another form of male resembles the female in certain particulars : *vide* Bateson and Brindley, Variation in Secondary Sexual Characters, P.Z.S., 1892, p. 585.

Micippa thalia, Herbst.

Cancer thalia, Herbst, Krabben, III. iii. 50, tab. lviii. fig. 3.
 ̖ *Micippa thalia*, Gerstäcker, Archiv. fur Naturgesch., XXII. 1856, p. 109; and Adams and White, 'Samarang' Crust., p. 15; and A. Milne-Edwards, Nouv. Archiv. du Mus., VIII. 1872, p. 238, pl. xi. fig. 1 ; and Kossmann, Reise Roth. Meer., Crust., p. 8 (*et varr.*); and MIERS, Zoology H. M. S. 'Alert,' pp. 182 & 198, and ANN. MAG. NAT. HIST., 1885, VOL. XV. p. 10 (*ubi synon.*), and 'Challenger' Brachyura, p. 70; and [Cano., Boll. Soc. Nat., Napol., III. 1889, p. 179]; and Ortmann, Zool. Jahrb. Syst., etc., VII. 1893, p. 60 ; and Henderson, Trans. Linn. Soc., Zool., 1893, p. 348.
Micippa thalia (= var. *aculeata*), de Haan, Faun. Japon. Crust., p. 98, pl. xxiii. fig. 3 ; and Krauss, Südafr. Crust., p. 51 ; and Bianconi, Mem. Ac., Bologna, III., 1851, p. 103, pl. x. fig. 2 ; and Kossmann, Reise Roth. Meer., Crust., pp. 5 and 8, pl. iii. fig. 5 ; and Hilgendorf, MB. Akad., Berl., 1878, p. 786 ; and Richters, Möbius, Meeresfauna, Maurit., p. 142 ; and Miers, Ann. Mag. Nat. Hist., 1885, Vol. XV. p. 11 (*ubi synon.*); and De Man, Journ. Linn. Soc., Zool., Vol. XXII. 1888, p. 20 ; and Mary J. Rathbun, Proc. U. S. Nat. Mus., Vol. XVI. 1893, p. 92.

97

Micippe miliaris, Gerstäcker, Archiv. fur Naturges., XXII. 1856, p. 110; and Heller, Crust. Roth. Meer., SB. Ak., Wien, XLIII. 1861, p. 298, pl. i. fig. 1; and Kossmann, Reise Roth. Meer., Crust., pp. 4 and 8; and Miers, Ann. Mag. Nat. Hist., 1885, Vol. XV., p. 11.

Micippa haanii, Stimpson, Proc. Ac. Nat. Sci., Philad., 1857, p. 217; and Miers, Zool. H. M. S. 'Alert,' pp. 517 and 524; and C. W. S. Aurivillius, Kongl. Sv. Vet. Ak. Handl., XXIII. 1888–89, No. 4, p. 52, pl. iv. figs. 1, 1*a*; and de Man, J. L. S., Zool., Vol. XXII. 1888, p. 20.

Micippe pusilla, Bianconi, Mem. Ac. Sci., Bologna, 1869, Vol. IX. p. 205, pl. i. fig. 1 : and Hilgendorf, MB. Ak., Berl., 1878, p. 787.

Micippa inermis, Haswell, P. L. S., N. S. Wales, Vol. IV. 1879, p. 445, pl. xxvi. fig. 3, and Cat. Austral. Crust., p. 24.

Body and ambulatory legs covered with a woolly tomentum.

Carapace with the regions fairly well-defined, the hepatic regions depressed, and the surface closely and evenly granular. From the granular surface there usually, but not always, arise several large vertical spines, which are typically disposed as follows : — one on either supra-ocular hood, two on the gastric region in the middle line, and two placed obliquely on either branchial region. Any or all of these spines may be suppressed. The lateral margins are armed with an irregular series of spines or spinules, and a few spinules may exist on the posterior border in the middle line.

The rostrum is deflexed nearly vertically in the adult female, less vertically in the adult male, and at an angle of 45° or less in the young male : it ends in two curved divergent spines.

The basal antennal joint is produced at its antero-external angle to assist in the formation of the floor of the orbit, but there is a wide hiatus between this process and the post-ocular spine, so that the floor of the orbit is incomplete.

The chelipeds in the adult male are as long as the carapace, are not much stouter than the other legs, and have slender palms, and long slender fingers which, though nearly straight, are closely apposable only in their distal half. In the adult female the chelipeds are equal in length to the post-orbital portion of the carapace, are slenderer than the other legs, and have tapering palms and minute fingers. The merus and carpus of the ambulatory legs are sometimes swollen.

In the Museum collection are specimens, representing all the varieties of this species, from Mergui, Burma, Orissa and Malabar, as well as from Hongkong and Nagasaki.

This species shows quite as well as *M. cristata* the close relation of *Micippa* to *Maia*.

Micippa margaritifera, Henderson.

Micippa margaritifera, Henderson, Trans. Linn. Soc., Zool., 1893, p. 348, pl. xxxvi. figs. 5–7.

Carapace symmetrically sculptured, closely crisply and finely granular, and with the hepatic regions deeply excavate : there are three coarse spinules, disposed in a triangle base outwards, on either branchial region, and a denticle at the anterior boundary of the branchial region ; and on the posterior margin are three smooth polished globules " exactly resembling pearls " inset.

The rostrum is long, vertically deflexed in both sexes, and incurved at the tip, which ends in two shallow lobes — the outer angle of each lobe being marked by a spinule.

The basal antennal joint has its antero-external portion greatly produced to complete the floor of the orbit.

The chelipeds in the male are a little longer than the carapace, and have the palms broadened and inflated, and the fingers closely apposable only at the tip. In the female the chelipeds are very much slenderer than the other legs, are only as long as the post-orbital portion of the carapace, and have the hand very slender and tapering. The ambulatory legs are remarkable for their large obtriangular foliaceous meropodites, which in the first pair are specially remarkable, as they are closely apposable to the front, to form, as in *Calappa,* a shield.

In the Museum collection are specimens from both sexes from the Andamans, from Ceylon (34 fms.), and from the Maldives (20–30 fms.).

Micippa margaritifera, var. *parca* nov. I distinguish, provisionally, as a variety, two ovigerous females from the Andamans, in which the middle " pearl " on the posterior border is replaced by a group of spinules, and in which the meropodites of the ambulatory legs are even more broadly foliaceous.

CYPHOCARCINUS, A. M.-Edw.

Cyphocarcinus, A. Milne-Edwards, Nouv. Archiv. du Mus., IV. 1868, p. 73 ; and Miers, Journ. Linn. Soc., Zool., XIV. 1879, p. 664.

Carapace elongate, subcylindrical, with the gastric region greatly elevated ; *the anterior part of the gastric region, along with the front, being vertically deflexed.* The rostrum is formed of two little horns, *each* of which is sharply bifurcate at the tip, one branch being directed forwards and outwards, the other being recurved upwards. The eyes are small and are sunk in small tubular orbits formed in the typical Periceroid manner. The antennæ are small : the basal joint has its antero external angle separated from the rest of the joint by a deep cleft. The external

99

254 *Carcinological Fauna of India.*

maxillipeds have the merus dilated at both the internal and external anterior angles. The chelipeds in the female are not longer than the 2nd pair of legs and are hardly stouter. The ambulatory legs have the dactylus recurved, strongly spinate along the posterior edge — prehensile. The sternum in the female forms a hollow, the mouth of which is completely closed by the broad and perfectly flat abdomen.

? *Cyphocarcinus minutus*, A. M.-Edw.

Cyphocarcinus minutus, A. Milne-Edwards, *loc. cit.* pl. xix. figs. 7–12.

Carapace elongate, subcylindrical, the lateral borders nearly parallel in their posterior two-thirds, gently convergent anteriorly. Besides the greatly elevated and anteriorly deflexed gastric region, there are two or three slight bulgings on the side of either branchial region, a slight elevation on the cardiac region, and a median prolongation — overlapping the abdomen — of the posterior border. The hepatic regions are very small and are not visible from the dorsal aspect. The supra-orbital border bears one or two little teeth. The second joint of the antennal peduncle is much enlarged, the third is clavate, and the flagellum is hardly to be distinguished from the hairs on the third joint. The chelipeds in the female are smooth, but the legs are hairy and have the joints, especially the merus, somewhat broadened. Two adult females, one from the Pedro Shoal, the other from the Andamans, are in the Museum collection. The larger of the two is 10 millim. long and has the carapace deeply encrusted by a colony of calcareous Polyzoa.

MACROCŒLOMA, Miers.

Macrocœloma, Miers, Journ. Linn. Soc., Zool., Vol. XIV. 1879, p. 665; and 'Challenger' Brachyura, p. 79.

Entomonyx, Miers, Zoology H. M. S. 'Alert,' p. 525.

Carapace subpyriform, but broadened anteriorly by the projecting orbits: the dorsal surface unarmed, or tuberculated, or with a few long spines: the margins without a series of elongated lateral spines, but often with a strongly developed lateral epibranchial spine, preceded by some smaller spines. The spines of the rostrum are well developed. The eyes are retractile within roomy projecting tubular orbits, which are formed much as in *Micippa*.

The antennæ have the basal joint considerably enlarged and armed distally with one or two spines. The mobile portion of the antenna is sometimes concealed by the rostrum, sometimes exposed. The merus of the external maxillipeds is broader than the ischium, and notched at the internal angle for the insertion of the palp.

100

The chelipeds in the male have the palms enlarged, and the fingers either arched and meeting only at the tip, or not. The ambulatory legs are rather short.

This genus might, without any unnatural stretch, be included with *Micippoides*, A. M.-Edw. (Journ. Mus. Godeffr. I., Crust., p. 254).

Macrocoeloma nummifer, n. sp., Plate IV. fig. 4.

Closely allied to *Macrocoeloma concava*, Miers, ' Challenger ' Brachyura, p. 81, pl. x. fig. 2 ; and to *Entomonyx spinosus*, Miers, Zoology H. M. S. 'Alert,' p. 526, pl. xlvii. fig. B.

Carapace rather more than ¼ longer than broad, with the regions well-defined : its surface is regularly and sharply tubercular and is armed with two sharp spines—one behind the other—on the gastric region, two larger—side by side—on the cardiac region, two still larger—one obliquely behind the other—on the lateral epibranchial region, and two very small ones—one behind the other—on the intestinal region.

The rostrum consists of two straight sharp slightly diverging spines, which are about one-fifth or one-sixth the length of the carapace proper, and which in the male are slightly deflexed, but in the female are strongly deflexed.

The basal joint of the antennæ is broadly obtriangular ; its antero-external angle is produced to aid in forming the floor of the orbit—this orbital process having its free margin deeply excised ; its antero-internal angle carries a stout vertically directed tooth. The orbits, which are in the form of large deep projecting tubes with jagged lips, are constituted as in *Micippa*.

The chelipeds are closely and sharply granular as far as the fingers: in the male they are much stouter than the other legs, are nearly as long as the carapace and rostrum, and have large broad palms, and strongly arched fingers that meet only at the tip. In the female the chelipeds, although not much shorter than those of the male, are hardly stouter than the other legs, and have fingers that can be closely apposed throughout their extent.

The ambulatory legs are slender: in all the meropodite has its posterior margin minutely spinulose, and has a spine on the far end of the upper margin : the first pair, which are the longest, are a little longer than the chelipeds.

The rostrum carapace and legs are beset with stiff curly hairs.

The abdomen in both sexes consists of seven distinct segments.

This species commonly encrusts itself with a plate armour of Orbitolites, rounded fragments of Nullipore, &c.

101

Loc. Andaman Sea, 17–36 fms.   Off Ceylon 34 fms.

		Male.	Adult female.
Greatest length	...	... 21 millim.	21 millim.
„ breadth	...	... 14 „	16 „
Length of chelipeds	...	... 19 „	15 „

TIARINIA, Dana.

Tiarinia, Dana, U. S. Expl. Exp., Crust., pt. I. p. 109.
Tiarinia, Miers, Journ. Linn. Soc., Zool., Vol. XIV. 1879, p. 664.

Carapace subpyriform, somewhat broadened anteriorly, tuberculated, terminating in a rostrum composed of two moderately deflexed horns which are in close contact with one another, except sometimes at the extreme tip.

The eyes are enclosed in tubular orbits formed by a prominent supra-ocular roof the anterior angle of which is strongly produced forwards, by a cupped post-ocular tooth, and by a process of the broad basal antennal joint, all three elements being in the closest contact. The mobile portion of the antenna is completely exposed.

The external maxillipeds have the merus broader than the ischium owing to the expansion of its external angle, and the palp inserted in a slight notch in the internal angle of the merus.

The chelipeds are little enlarged in the male : the ambulatory legs have the dactylus short and claw-like.

The abdomen in both sexes consists of seven distinct segments.

Tiarinia cornigera, (Latr., Edw.)

[*Pisa cornigera,* Latr., Encyc., X. 141.]
Pericera cornigera, Milne-Edwards, Hist. Nat. Crust., I. 335 ; and Adams and White, ' Samarang ' Crust., p. 18.
Tiarinia cornigera, Dana, U. S. Expl. Exped., Crust., pt. I. p. 110, pl. iii. figs. 5*a-e* ; and Stimpson, Proc. Acad. Nat. Sci., Philad., 1857, p. 217 ; and Haswell, Proc. Linn. Soc., N. S. Wales, Vol. IV. 1879, p. 449, and Cat. Austral. Crust., p. 28 ; and Miers, Ann. Mag. Nat. Hist., 1880, Vol. V. p. 228 ; and Mary J. Rathbun, Proc. U. S. Nat. Mus., Vol. XV. 1892, pp. 243 and 276.
? *Pericera tiarata* and *setigera,* Adams and White, ' Samarang ' Crust., p. 17.
Tiarinia verrucosa, Heller, ' Novara ' Crust., p. 4, taf. i. fig. 3.
Tiarinia mammillata, Haswell, Proc. Linn. Soc., N. S. Wales, Vol. IV. 1879, p. 448, and Cat. Austral. Crust., p. 27.

Body and ambulatory legs with many curly hairs.

Carapace pyriform, the regions well-defined, the surface closely and very variedly pustular nodular and granular, but with the following markings fairly constant :—two parallel longitudinal lines of small nodules between the orbits ; a " cross " of larger nodules on the gastric region, the base of the cross being formed by three pustules; three pustules arranged in a triangle base forwards on the cardiac region, behind which are three conical tubercles arranged in a transverse line; a coarse claw-like tooth at the lateral epibrancial angle.

The rostrum consists of two moderately deflexed spines, which are parallel, and in the closest contact, either throughout their extent, or to near the tips, which may then be upcurved and slightly divergent: the length of the rostrum varies from nearly one-half to one-fourth the length of the carapace, its usual length is about ⅔ths that of the carapace.

The antennæ have the basal joint broadened and produced to form the floor of the orbit, the antero-external angle being further produced to form a coarse spine: the next two joints are broadened and fringed with stiff bristles: the flagellum is short. The eyes are ensheathed in orbits which are formed as already described: the supra-ocular eave has a dog's-ear form, and the post-ocular tooth is also salient. The chelipeds in the adult male are as long as the carapace without the rostral spines, and are a little stouter than the other legs: the merus is nodular, most markedly so on the upper surface; the carpus is granular; and the palm—which is a good deal broadened and inflated—and the fingers, are smooth and polished, the fingers being arched and meeting only at tip.

In the female and young male the chelipeds are only as long as the post-orbital portion of the carapace, are slenderer than the other legs, and have the palm slender, the fingers however being arched.

The ambulatory legs are stout, and have strong claw-like dactyli, the posterior border of which is denticulate; the ischium in all is swollen, and is more or less nodular on the upper surface; and the carpus in all is broadened: the first pair, which are considerably the longest, slightly exceed the length of the carapace and rostrum.

In the Museum collection are forty well preserved specimens from the Andamans.

The closeness of the relation between *Tiarinia* and *Micippa* is well seen in the very young of the above species, in which the carapace is depressed and is so broad in front as to be almost oblong, and the rostrum is deflexed at an angle of 45°.

Family II. PARTHENOPIDÆ.

Parthenopiens (part.) and *Canceriens cryptopodes,* Milne-Edwards, Hist. Nat., Crust., I. pp. 347 and 368.

Parthenopinea, Dana, U. S. Expl. Exp., Crust., I. pp. 77 and 136.

Parthenopinea, Miers, Journ. Linn. Soc., Zool., Vol. XIV. p. 641; and 'Challenger' Brachyura, p. 91.

The eyes are usually retractile within small circular well-defined orbits, the floor of which is nearly continued to the front, leaving a hiatus which is usually filled by the second joint of the antennary peduncle. The basal antennal joint is small, and is deeply imbedded between the inner angle of the orbit and the antennulary fossæ.

The antennules fold a little obliquely.

The *Parthenopidæ* are divided by Miers into two sub-families, namely:—

Sub-family I. *Parthenopinæ*; in which the carapace is sometimes sub-pentagonal or ovate-pentagonal, more commonly equilaterally-triangular, and sometimes almost semi-circular or semi-elliptical in outline; in which the cardiac and gastric regions are usually so deeply marked off from the branchial regions on either side as to make the dorsal surface of the carapace trilobed; in which the chelipeds are vastly longer and more massive than the ambulatory legs; and in which the rostrum is either simple or obscurely trilobed.

Sub-family II. *Eumedoninæ*; in which the carapace is, commonly, sharply pentagonal, with the junction of the antero-lateral and postero-lateral borders strongly produced; in which the cardiac and gastric regions are not conspicuously marked off from the branchial regions; and in which the chelipeds are of moderate size.

Sub-family I. PARTHENOPINÆ, Miers.

Miers, Journ. Linn. Soc., Zool., Vol. XIV. 1879, p. 668.

Key to the Indian genera.

I. Carapace not laterally expanded:—
　　1. Basal antennal joint very short, not nearly reaching the inner canthus of the orbit: fingers of chelipeds very strongly incurved... LAMBRUS.
　　2. Basal antennal joint nearly reaching the inner canthus of the orbit: fingers slightly incurved .... PARTHENOPE.

II. Carapace more or less expanded to form a vault in which the ambulatory legs are concealed:—
　　1. Carapace transversely triangular; greatly expanded both laterally and posteriorly ...... CRYPTOPODIA.
　　2. Carapace transversely triangular; expanded laterally, but not posteriorly: a ridge on the pterygostomian region................................ HETEROCRYPTA.

3. Carapace transversely oval; expanded laterally, but not posteriorly : no ridge on the pterygostomian region.............. . ... ŒTHRA.

LAMBRUS, Leach.

Lambrus, Leach, Trans. Linn. Soc., Vol. XI. 1815, pp. 308, 310.
Lambrus, Milne-Edwards, Hist. Nat. Crust., I. 352.
Lambrus, A. Milne-Edwards, Miss. Sci. Mex., Crust., I. p. 146.
Lambrus, Miers, J. L. S., Zool., Vol. XIV. 1879, p. 668; and ' Challenger ' Brachyura, p. 91.

Carapace either broadly triangular with rounded sides and pointed front, or ovate-pentagonal with front pointed but extremely short: the surface is granular, or tubercular, or spiny.

The eyes are enclosed in distinct orbits, which have a suture above and a hiatus below, the hiatus being occupied by the second joint (true third joint) of the antennal peduncle.

The antennules fold obliquely. The antennæ are small: their basal joint, which is extremely short, and does not reach the front, is wedged in between the antennulary fossa and the large lobe that constitutes the floor of the orbit.

The buccal frame is usually quadrangular, but is sometimes a little narrowed in front; it is completely closed by the external maxillipeds : the epistome is sometimes very large, sometimes narrow.

The chelipeds are usually of immense size and length, out of all proportion to the short slender ambulatory legs : the meropodite and " hand " are usually prismatic, with the borders strongly dentate: the fingers are much shorter than the palm, and are abruptly curved inwards and a little downwards.

The abdomen of the female usually consists of seven segments; that of the male of five or six.

Professor A. Milne-Edwards, (Miss. Sci. Mex., Crust., I. pp. 146–148) subdivides the genus Lambrus into ten sub-genera, the independence of all of which, however, is not universally admitted.

The sub-genera at present known to exist in Indian waters are shown in the following

Key to the Indian sub-genera of the genus Lambrus.

I. Carapace tuberculate, ovate-pentagonal, the rostrum not breaking beyond the general outline of the body: the buccal frame a little narrowed in front.................... LAMBRUS.

105

II.   Carapace strongly carinated or tuberculated, broadly triangular (considerably broader than long), with rounded
sides and a broad but sharp-pointed projecting rostrum : no post-ocular constriction : chelipeds with the
arm and hand straight, sharply trigonal, the edges of
these joints, as also the outer edge of the carpus, being
very sharply and stoutly serrated.............................. PLATYLAMBRUS.

III. Carapace granular or spiny, usually as long as broad, with
a projecting rostrum, and a very distinct post-ocular
constriction .... RHINOLAMBRUS.

IV.   Carapace granular, broader than long, and with the postero-
lateral angle produced to form a great blade-like spine.
Pterygostomian region deeply channelled, obliquely, the
channel being closed below by thick fringes of hairs .... AULACOLAMBRUS.

V.   Carapace worn and eroded, broader than long, almost semi-
circular in outline, with the postero-lateral angle produced; the rostrum more or less deflexed, and not, or
hardly, breaking the general outline : no post-ocular,
but a fairly distinct post-hepatic constriction : chelipeds with the arm and hand indefinitely contorted, not
sharply trigonal ; and with their edges, if spinate, irregularly and bluntly so ; the carpus quite smooth exter-
nally : the chelipeds are short for the genus................ PARTHENOLAMBRUS.

Sub-genus LAMBRUS, A. Milne-Edwards.

Lambrus, A Milne-Edwards, Miss. Sci. Mex. Crust., I. p. 146.
Lambrus, Miers, 'Challenger' Brachyura, p. 92, (*part.*)

Carapace ovate-pentagonal, with the surface granular or pustular
and but little carinate in the adult : rostrum exceedingly short.

Lambrus longimanus, Leach.

? Cancer spinosus longimanus, Rumph, Amboin. Rariteitk., pl. viii. fig. 2.
Cancer macrochelos, Seba, III. xix. 1, 8, 9.
? Parthenope longimanus, Fabr. Suppl., p. 353.
? Cancer longimanus, Linn., Syst. Nat., II. 1016, 42.
? Cancer longimanus, Herbst, Krabben, I. ii. 253, taf. xix. figs. 105, 107.
  Lambrus longimanus, Leach, Trans. Linn. Soc., Vol. XI. 1815, p. 310 ; and
Milne-Edwards, Hist. Nat. Crust., I. 354 ; and Cuvier, Regne Animal, pl. xxvi. fig.
1 ; (and *? Lambrus longimanus*, Adams and White, 'Samarang' Crust., p. 30) ; and
Bleeker, Crust de l'Ind. Archip., p. 17 (*nec syn.* pelagicus, *Rupp.*) ; and Miers, Ann.
Mag. Nat. Hist., 1879, Vol. IV. p. 20, and Zoology H. M. S. 'Alert,' pp. 182 and 200,
and 'Challenger' Brachyura, p. 95 ; and W. A. Haswell, P. L. S., N. S. Wales, Vol.
IV. 1879, p. 449, and Cat. Austral. Crust., p. 31 ; and A. O. Walker, J. L. S., Zool.,
Vol. XX. 1890, p. 109 ; and de Man. J. L. S., Zool., Vol. XXII. 1888, p. 21 (*ubi
synon.*) ; and Henderson, Tr. Linn. Soc., Zool., (2) V. 1893, p. 349.

Carapace almost oval transversely, and with the surface granular or pustular. (In the young, besides tubercles, there are some coarse spinules in five series—a median, and two oblique lateral on either side.) The lateral borders are spinulate or crenulate anteriorly, spinate posteriorly, smooth quite posteriorly at the junction with the posterior border: the posterior border, except for a hook-like spinule at either end, and two spinules in the middle line, is smooth: there are often one or two curved spines on the branchial region : the pterygostomian region is quite smooth, but on the inferior branchial region are a few coarse spinules, most distinct at the bases of the legs.

The rostrum, which is symmetrically trilobed, is very small, its length being less than one-twelfth that of the rest of the carapace.

The chelipeds, which are massive, are about four times the length of the carapace in the male, about $3\frac{1}{2}$ times in the female : the *meropodite* is prismatic, or, in transverse section, rhomboidal ; its anterior and posterior edges are armed with numerous, somewhat curved, spines —alternating larger and smaller; its upper edge, as sometimes either upper surface, has a row of spinules ; its lower edge is rounded, and has a discontinuous series of spinules ; its under surfaces are smooth and polished : the *carpus* has 3 or 4 sharp thin teeth on its outer margin : the trigonal *palm* has twelve or more sharp thin laciniated teeth on its outer edge—alternately larger and smaller; along its inner edge is a long series of multicuspid spines ; its under edge is finely beaded, and its under surfaces are almost smooth; its upper surface has numerous irregularly disposed spinules and granules : the *dactylus* has numerous spinules on the outer surface of its broad base.

The ambulatory legs have the merus compressed and spinulate as to its edges, especially the posterior (inferior) edge : the longest of the ambulatory legs is hardly longer than the meropodite of the chelipeds.

Colours in life, pale lilac dorsally, white ventrally.

In the Museum collection are numerous specimens from the Madras coast, from Arrakan and Mergui, and from the Andamans.

Sub-genus PLATYLAMBRUS, Stimpson.

Platylambrus and *Enoplolambrus*, A. Milne-Edwards, Miss. Sci. Mex., Crust., I. pp. 146 and 147.

Lambrus, Miers, ' Challenger' Brachyura, p. 92 *(part)*.

Carapace carinated or tuberculated, broader than long, broadly triangular with rounded sides and a broad but acute and projecting rostrum : no post-ocular constriction : chelipeds with the meropodite dan palm straight, the former joint prismatic, the latter sharply tri-

107

gonal, the anterior and posterior borders of both joints sharply laciniate or serrate, as is also the outer edge of the carpus.

Key to the Indian species of the sub-genus Platylambrus.

I. Carapace with three distinct carinæ, one median, and one, oblique, on either side: chelipeds with their surfaces (but not their edges) for the most part smooth: ambulatory legs, with few spines.	1. Infra-orbital lobe entire and strongly produced at the inner (inferior) angle to form a great spine plainly visible from above on either side of the rostrum .... *L. prensor.*
	2. Infra-orbital lobe deeply cleft, the inner portion not or hardly visible from above *L. carinatus, Edw.*

II. Carapace covered with great mushroom-like or paxilliform tubercles: chelipeds with their surfaces very strongly spinate or tuberculate: ambulatory legs strongly spiniferous.. *L. echinatus.*

Lambrus (Platylambrus) prensor, Herbst.

Lambrus prensor, Herbst, Krabben, II. ii. 170, tab. xli. fig. 3.
Lambrus prensor, Milne-Edwards, Hist. Nat. Crust., I. 358.
Lambrus jourdainii, F. de B. Capello, Jorn. Sci. Lisb., III. 1870-71, tab. 3, fig. 6.
Lambrus prensor, A. Milne-Edwards, Nouv. Archiv. du Mus., Vol. VIII. 1872, p. 260 (foot-note) ; and Miss. Sci. Mex., Crust., I. p. 147 (foot-note).
Lambrus prensor, Walker, J. L. S. Zool., Vol. XX. 1890, p. 109 (name only).

Our numerous specimens correspond exactly with Capello's figure and succint and graphic description. M. A. Milne-Edwards at first assigned Capello's species to *L. carinatus,* Edw., but afterwards to *L. prensor,* and it is this last authority that I now follow.

Carapace broader than long, broadly triangular with the sides rounded: the median and branchial regions are strongly prominent, the former having three small spinules in the middle line, the latter having each two oblique granular ridges, one of which is very faint and runs to the large lateral epibranchial spine, the other of which forms a strong carina, and runs to the large spine at the postero-lateral angle. The antero-lateral margin is armed with 7 or 8 nearly equal-sized close-set compressed teeth, behind which, at the lateral epibranchial angle, is a very large blade-like spine: behind this again, on the postero-lateral border are two large teeth, the outer of which, at the postero-lateral angle, is nearly as large as the lateral epibranchial spine ; and lastly on the posterior border are three large curved spines.

The rostrum is acute, concave at base, and slightly recurved at tip : on either side of the rostrum is seen from above a very strong and acute spine formed by the prolongation of the inner margin of the infra-orbital lobe—this lobe is entire.

The chelipeds are massive and are about three times the greatest length of the carapace : their surfaces are almost smooth : the arm is rhomboidal in transverse section, and the palm is sharply trigonal : the lower edges of the arm, wrist and palm form a continuous line of beading : the upper edge of the arm is granular and spinular : the inner or anterior edges of the arm, wrist and hand are spinate — the spines growing larger towards the end of the palm, while the posterior (or outer) edges of the same three joints are very strongly and closely laciniate.

As usual the spines in all cases have a tendency to be alternately larger and smaller.

Of the ambulatory legs the merus, carpus and propodus have the anterior (upper) border strongly and sharply carinate, while the merus has also the posterior border spinate.

This species is not uncommon along the Orissa coast, from 8 to 23 fathoms.

Lambrus (Platylambrus) carinatus, Edw.

Lambrus carinatus, Milne-Edwards, Hist. Nat. Crust., I. 358.
Lambrus carinatus, A. Milne-Edwards, Miss. Sci. Mex., Crust., I. p. 147 (foot-note).

Our specimens, which agree with the diagnoses of M. A. Milne-Edwards completely, are distinguished from those above described as *L. prensor*, (1) by having the mid-dorsal carina formed by three great compressed teeth ; (2) by the single, and very high and sharply cut carina on either branchial region ; (3) by the smaller size of the spine at the lateral epibranchial angle and of the spine, at the postero-lateral angle, immediately succeeding it ; (4) by the form of the infra-orbital lobe, which instead of being entire, is bilobed — the inner lobe, more-over, having a rounded apex, and not being visible from above ; (5) by the meropodites of the ambulatory legs having their anterior (upper) edge serrate, not carinate, and by the carpopodites and propodites having the anterior edge smooth.

These differences are constant in a series of twelve specimens, including both sexes.

This species also differs from *L. prensor* in its much smaller size, three ovigerous females having the carapace 11 millim. in its greatest breadth (exclusive of spines), while ovigerous females of *L. prensor* have the carapace 28 to 30 millim. in its greatest breadth exclusive of spines.

[? *Lambrus* (*Platylambrus*) *holdsworthii*, Miers.

Lambrus holdsworthii, Miers, Ann. Mag. Nat. Hist., Vol. IV. 1879, p. 19, pl. v.
fig. 3; and 'Challenger' Brachyura, p. 93 and Henderson, Trans. Linn. Soc., (2) V.
1893, p. 350.

The single specimen that I doubtfully refer, from Miers' figure and
description, to this species, has a close resemblance to both the species
identified above as *L. prensor* and *L. carinatus*. It differs from them
both (1) in having numerous scattered tubercles on the carapace, and
(2) in having the large spine at the lateral epibranchial angle and the
two outer spines on the postero-lateral margin all of about the same
size. It resembles *L. prensor*, and differs from *L. carinatus*, in not
having the branchial region traversed by a single sharp-cut carina :
and it resembles *L. carinatus*, and differs from *L. prensor*, in having a
median line (though not a high carina) of three large teeth, in having
the infra-orbital lobe deeply cleft and not exceedingly produced, and in
having the anterior (or upper) edge of the meropodites of the ambulatory legs dentate instead of carinate.]

Lambrus (*Platylambrus*) *echinatus*, Herbst.

Cancer echinatus, Herbst, Krabben, I. ii. 255, taf. xix. figs. 108-109.
Parthenope giraffa, Fabr., Supplement, p. 353.
[*Maia echinatus* and *giraffa*, Bosc, I. 250].
Lambrus giraffa, Desmarest, Consid. Crust., p. 85.
Lambrus echinatus, Milne-Edwards, Hist. Nat. Crust., I. 356.
Lambrus echinatus, Miers, 'Challenger' Brachyura, p. 93.

Carapace broader than long, broadly triangular with the sides
rounded : the gastric and cardiac regions are elevated, and are delimited
on either side from the elevated branchial regions by broad and deep
grooves. The entire carapace is covered, but not very densely, with
large mushroom-like and paxilliform tubercles, the spaces between
which are occupied, but not densely, by short, crisp, upstanding hairs.
The lateral margins are armed with ramose spines, which increase in
size from before backwards : the posterior and part of the postero-
lateral margins are armed with tubercles like those on the surface of
the carapace. The granular rostrum is broad and concave at the base,
and is then suddenly narrowed to form a little peak.

The chelipeds which are from 3½ (female) to 3¾ (male) the greatest
length of the carapace, are distinguished by having their upper aspect
(edges and surfaces) covered with ramose spines, and their under aspect
covered with great pearly tubercles. The ambulatory legs are distin-

gnished by the large and numerous spines on their 3rd, 4th and 5th joints.

This species is not uncommon off the Orissa coast from 7 to 23 fathoms.

Sub-genus RHINOLAMBRUS, A. Milne-Edwards.

Rhinolambrus, A. Milne-Edwards, Miss. Sci. Mex., Crust., I. p. 148.
Lambrus, Micrs, ' Challenger ' Brachyura, p. 92 (*part.*).

Carapace triangular, usually as long as broad, with a broad projecting somewhat declivous rostrum and a very distinct post-ocular constriction; surface of carapace very commonly, but not always, spiny and granular.

Key to the Indian species of the sub-genus Rhinolambrus.

I. Chelipeds stout, three times to twice or less the length of the carapace and rostrum.

  1. Carapace and chelipeds very closely covered with large rugged granules and sharp ramose spines.
- i. Chelipeds nearly three times the length of the carapace and rostrum........ *L. contrarius.*
- ii. Chelipeds not two-and-a-half times the length of the carapace and rostrum........ *L. longispinis.*

  2. Carapace with few depressed tubercles, or nearly smooth: chelipeds with blunt teeth or smooth granules.
- i. Chelipeds three times the length of the carapace and rostrum........ *L. pelagicus.*
- ii. Chelipeds not twice the length of the carapace and rostrum .... *L. gracilis.*

II. Chelipeds slender, three-and-a-half to five times the length of the carapace and rostrum.

  1. Carapace at least as long as broad: large erect turret-like spines on the carapace.
- i. A single turret on the cardiac region, and on either branchial region: two large diverging spines in the middle line on the posterior border... *L. turriger.*
- ii. Two turrets on the cardiac region, and two on either branchial region: a single spinule on the posterior margin............. *L. cybelis.*

  2. Carapace broader than long; large spines of ordinary form on the carapace *L. petalophorus.*

111

Lambrus (Rhinolambrus) contrarius, Herbst.

Cancer contrarius, Herbst, Krabben, III. iv. 18, tab. lx. fig. 3.
[*Parthenope spinimana,* Lamk., Hist. Anim. Sans. Vert., V. 239.]
Lambrus spinimanus, Desmarest, Consid. Crust., p. 86, pl. iii. fig. 1.
Lambrus contrarius, Milne-Edwards, Hist. Nat. Crust., I. 354.
Lambrus contrarius, Bleeker, Recherches Crust. de l' Ind. Archip., p. 18.
Lambrus contrarius, A. Milne-Edwards, Maillard's l' ile Réunion, Annexe F, p. 10.
Lambrus contrarius, Brocchi, Ann. Sci. Nat., (6) II. 1875, Art. 2, p. 98, pl. xviii.
figs. 166, 167 (♂ appendages).
Lambrus contrarius, Richters, in Möbius, Meeresf. Maurit., p. 145.
Lambrus contrarius, Miers, Ann. Mag. Nat. Hist., 1880, Vol. V. p. 230; and
'Challenger' Brachyura, p. 94.
Lambrus contrarius, J. R. Henderson, Trans. Linn. Soc., Zool., (2) V. 1893,
p. 350.

Carapace, with rostrum, slightly longer than broad, everywhere covered with jagged granules and spines : the regions are strongly convex, and, usually, in the middle line, are three or four, and again on either branchial region, one or two spines of predominant size. The rostrum is broad, prominent, declivous, and spiny or granular, both on the upper surface and along the margins. The hepatic regions are very prominent, and their angle is strongly produced. The orbital edge is prominent and the post-orbital constriction strongly pronounced.

The chelipeds are about three times the length of the carapace and rostrum, and are extremely massive, the hands especially : above they are covered with large sharp jagged spines with rough tubercles interpersed ; below they are everywhere covered with rasp-like granules, The ambulatory legs are rather stout for a *Lambrus,* and have the merus somewhat spiny along one or both edges.

Colours in spirit, mottled pink, tips of fingers purple-black, ambulatory legs banded alternately yellow and bluish pink.

Our largest specimens, a male and a female, are from off Colombo, 26½ fathoms, and have a span (of chelipeds) of 290 millim. and 265 millim. respectively.

Lambrus (Rhinolambrus) longispinis, Miers.

Lambrus longispinus, Miers, Ann. Mag. Nat. Hist., 1879, Vol. IV. p. 18; Zoology II. M. S. 'Alert,' pp. 182 and 199; and 'Challenger' Brachyura, p. 93.
Lambrus longispinus, de Man, Archiv. fur Naturgesch., LIII. 1887, p. 229.
Lambrus longispinus, Walker, Journ. Linn. Soc., Zool., Vol. XX. 1890, p. 109.
Lambrus longispinus, Henderson, Trans. Linn. Soc., Zool., (2) V. 1893, p. 350.
Lambrus spinifer, Haswell, P. L. S., N. S. Wales, Vol. IV. 1879, p. 451, pl. xxvii. fig. 1; and Cat. Aust. Crust., p. 34.

Carapace, with rostrum, little longer than broad, its surface covered with spiny tubercles : There are four prominent spines in the middle

line, of which three are on the cardiac and one is on the gastric region ;
in front of the latter are two smaller spines placed transversely : on
the branchial regions are some small spines set in two oblique series,
and one large spine. On the antero-lateral margins are about nine
small close-set blunt faintly-laciniated teeth, slightly increasing in size
posteriorly ; on the postero-lateral margin are two large spines ; and
on the posterior border, in the middle line, is a pair of spines. The
rostrum is broad, prominent, acute and declivous. The post-ocular
constriction is distinct ; and the hepatic regions are well marked, with
the outer border denticulate. The chelipeds in the male are about 2¼
times the length of the carapace and rostrum : they much resemble
those of *L. contrarius*, the spines being for the most part jagged, and
the tubercles rasp-like. On the anterior (inner) margin of the arm are
10 or 12 spines alternating in size, the last three being very small ;
on the upper surface of the arm three spines are very prominent, as
are three or four on the posterior (outer) edge. On the anterior
(inner) margin of the hand are 7 or 8 spines increasing in size from
behind forwards ; while on the posterior margin are numerous spines
—only three or four of which are large. The lower surface of the
arms, wrists and hands is closely covered with large round rasp-like
tubercles. The merus and sometimes the two following joints of the
ambulatory legs, have the margins dentate.

Our single specimen from the Arrakan coast, 13 fms., is plainly
the same as Haswell's *L. spinifer*, judging from his figure (*tom. cit.*)
Both from that figure and from our specimen I should consider the
species to be more nearly related to *L. contrarius* than to *L. validus*.

Lambrus (Rhinolambrus) pelagicus, Rüpp.

Lambrus pelagicus, Rüppell, Beschr. u. Abbild. 24 Art. Krabben des Roth.
Meer., p. 15, pl. iv. fig. 1.
Lambrus pelagicus, Milne-Edwards, Hist. Nat. Crust., I. 355.
Lambrus pelagicus, Rüpp. (prob. = *affinis*, A. M.-Edw.) Miers, Ann. Mag. Nat.
Hist., 1879, Vol. IV. p. 21.
Lambrus pelagicus, Ortmann, Zool. Forsch. in Austral. u. Malay. Archip., Jena,
1894, p. 46.
Lambrus affinis, A. M.-Edw., Nouv. Archiv. du Mus., VIII. 1872, p. 261, pl. xiv.
fig. 4.
Lambrus affinis, Haswell, Cat. Austral. Crust., p. 34.
Lambrus affinis, Miers, 'Challenger' Brachyura, p. 95.
Lambrus affinis, J. R. Henderson, Trans. Linn. Soc., Zool. (2) V. 1893, p. 350.
[*Lambrus affinis*, F. Muller, Verh. Ges. Basel, VIII. p. 473.]
[*Lambrus affinis*, Cano, Boll. Soc. Nat. Napol., III. 1889, p. 187.]

Carapace, with rostrum, as long as broad : its regions well delimited and faintly pitted and pimpled, the furrows between the regions
113

being smooth and bare—except for a pimple at each of the four angles of the cardiac region. On either branchial region, above the postero-lateral angle of the carapace, is a bluntly conical spine. The rostrum is very broad, and is concave and bluntly pointed: on either side above the eye is a little eminence which carries a tuft of long silky hairs. The post-ocular constriction is distinct, as is also the post-hepatic. The antero-lateral (including the hepatic) margin is faintly crenulated: the posterior border is quite smooth.

The chelipeds in the male are three times the length of the cara-pace, but not more than 2½ times in the female: the anterior (or inner) margin of the arm and hand is evenly and bluntly dentate, or crenulate; the posterior (or outer) margin in the same joints is as evenly but much more bluntly and indistinctly dentate, and the lower margin faintly beaded: the carpus is either quite smooth or has a few nodules.

The ambulatory legs are smooth, rather stout, and are longer than the hand. In the male near the anterior border of the 6th abdominal tergum is a strong spine. This is a fairly common species at the Andamans.

Lambrus (Rhinolambrus) gracilis, Dana.

Lambrus gracilis, Dana U. S. Expl. Exp. Crust., pt. I. p. 137, pl. vi. figs. 6 *a–b.*
Lambrus gracilis, Miers, ' Challenger ' Brachyura, p. 94.
Lambrus deflexifrons, Alcock and Anderson (*nec* Miers), J. A. S. B., 1894, pt. ii. p. 199.

Carapace, with rostrum, considerably longer than broad; with a pronounced post-ocular constriction; somewhat rhomboidal in shape: the regions are extremely prominent, especially the cardiac, which is capped by a conical tooth, and the branchial, which rises into an oblique crest terminating posteriorly in a tooth: the hepatic region forms a prominent tooth, behind which the rounded lateral margins are 6 or 7 toothed: there are two laminar teeth on the posterior border: other-wise the carapace is smooth. The rostrum is broad, deflexed, and dis-tinctly trilobed towards the tip.

The chelipeds are not quite twice the length of the carapace and rostrum; and in the adult are not symmetrical—one, either right or left, having the hand much larger than the other. In the young the asymmetry is hardly noticeable. The *arm* has the anterior (inner) and posterior (outer) border irregularly armed with compressed blunt spines, of which the one at the far end of the outer border is the largest —being almost foliaceous: the *hand* has its inner and outer borders armed in the same irregular way, two or three of the teeth on the outer border, and one on the inner border being enlarged: the under surfaces of the chelipeds are quite smooth, but the upper surface of the arm has an incomplete longitudinal line of beading. The ambulatory legs are long and particularly slender.

In the Museum collection are specimens of males, ovigerous females and young, from the Andamans and from off Ceylon.

Lambrus (Rhinolambrus) deflexifrons, Miers.

Lambrus deflexifrons, Miers, Ann. Mag. Nat. Hist., Vol. IV. 1879, p. 21, pl. v. fig. 5. Ceylon.

This species, which is not represented in the Museum collection, is described as follows by Miers :—

"The carapace is strongly constricted behind the orbits, with the cardiac region very convex, and with an oblique but shallow sulcus ou the branchial regions, and is covered with closely-set small tubercles; the antero-lateral margins are unarmed; but there are two larger tubercles or small spines on the postero-lateral margins. The rostrum is vertically deflexed, triangular, and granulated above. The basal antennal joint is very small; the epistoma is large; the sub-hepatic and pterygostomian regions are not channelled. The anterior legs have the arm rounded and tuberculate above, with small spines on its anterior margin; the wrist is tuberculate; the hand with a few tubercules on its upper surface, the anterior margin armed with about ten, and the posterior with four granulated spines. The under surface of arm, wrist, and hand is closely granulated. The ambulatory legs are smooth, and are not compressed and cristate as usual in the genus.

The vertically deflexed rostrum and carapace, devoid of spines ou its surface and anterior margins, and non-compressed ambulatory legs are characteristic of this species. It seems to be allied to *L. gracilis*, Dana, a species from the Fijis, in the form of the carapace and legs; but in that species the carapace has a spine on the cardiac and each branchial region, and elsewhere appears to be smooth."

Lambrus (Rhinolambrus) turriger, Ad. & Wh.

Lambrus turriger, White, P. Z. S., 1847, p. 58; Ann. Mag. Nat. Hist., Vol. XX. 1847, p. 63; and Adams and White, 'Samarang' Crust., p. 26, pl. v., fig. 2.

Lambrus turriger, W. A. Haswell, Proc. Linn. Soc., N. S. Wales, Vol. IV. 1879, p. 449; and Cat. Austral. Crust., p. 32.

LAMBRUS TURRIGER, MIERS, ZOOLOGY H. M. S. 'ALERT,' p. 201; and 'Challenger' Brachyura, p. 96.

Carapace, with rostrum, a little broader than long; slightly granular; the regions well-defined and armed with huge, erect or semi-erect, knob-headed spines, as follows :—one on the gastric region, in the mid-

115

dle line, one on the cardiac region in the middle line, and one on each branchial region : there is sometimes a little spinule in front of the gas-tric spine, and one in front of either branchial spine ; and on the pos-terior border, in the middle line, are two divergent spines directed back-wards. The rostrum is broad, concave between the eyes, somewhat de-flexed, and may be described as trilobed near the tip—since it is there suddenly truncated and continued in the middle line only.

There is a distinct post-ocular constriction, and the hepatic regions are well-defined laterally.

The chelipeds are long slender and rugose : the arm is cylindrical, and the palm subcylindrical, becoming enlarged and trigonal near the fingers : in the male the chelipeds are from $4\frac{1}{2}$ to $5\frac{1}{2}$ times the length of the carapace and rostrum, in the female they are but $3\frac{1}{3}$ to $3\frac{1}{2}$ times this length.

The ambulatory legs are long, very slender, and perfectly smooth.

In the Museum collection are numerous specimens from the Anda-mans, from the Madras coast, and from off Ceylon at 32 to 34 fathoms.

There are undoubtedly two sorts of males : one sort resembling the female in having the chelipeds comparatively short, the other sort hav-ing very long chelipeds.

Lambrus (Rhinolambrus) cybelis, n. sp.

This species closely resembles *L. turriger,* from which it differs only in the following characters :—

(1) the regions of the carapace are all more elevated, and on the cardiac region—one behind the other, in the middle line—as well as on either branchial region, are *two* very large semi-erect spines of equal size ; while in the middle of the granular posterior border is a single spinule :

(2) the surface of the carapace, besides being granular, is very evenly and regularly pitted or reticulated :

(3) the rostrum, which is nearly one-third the greatest breadth of the carapace, is more distinctly trilobed :

(4) the chelipeds (which in females and young males are only $3\frac{1}{4}$ to $3\frac{1}{2}$ times the length of the carapace and rostrum), though of the same general slender proportions,as in *L. turriger,* have the hand distinctly trigonal throughout, and the arm and hand armed with sharp laciniated spines on the upper aspect.

A young male from off Ceylon, 34 fms., and two probably half-grown males, and an ovigerous female, from off the Andamans, 41 to 86 fathoms.

116

The characters that distinguish this species are constant throughout the series, without any modification or variation.

Greatest length of carapace in ovigerous female ... 15 millim.
Do. breadth do. do. do. ... 15 millim.
Length of chelipeds in ovigerous female ... 52 millim.

Lambrus (Rhinolambrus) petalophorus, n. sp.

Carapace of the same general shape as in *L. turriger*, but broader posteriorly, where its breadth exceeds its length with the rostrum. The hepatic region is extremely well demarcated, not by its prominence, but by its almost vertical outer wall.

The cristiform antero-lateral border, which runs from the angle of the buccal frame outside the limit of the hepatic region, is festooned by 7 or 8 close-set thin teeth, and there is a strong upcurved spine at the postero-lateral angle.

The postero-lateral border carries three teeth, the innermost of which is hardly less prominent than that at the postero-lateral angle: the posterior border is finely denticulated.

The rostrum, the breadth of which is about ⅔ the greatest breadth of the carapace, is elegantly trilobed.

The regions of the carapace are strongly elevated, and have the surface pitted or reticulated: in the middle line on the gastric region is a single erect conical spine, on the cardiac region two; and on either branchial region there is a spine. In front of the gastric spine are two spinelets, disposed transversely.

The supra-orbital margin is strongly arched, and the infra-orbital lobe is cut into two elegantly crimped leaflets or petals.

The post-ocular constriction is distinct.

The chelipeds in the male are four and-a-half times the length of the carapace and rostrum : the *arm* is slender and subcylindrical, with a line of many spinules along both the inner and outer borders, a broken line of sharp tubercles along its upper surface, and a line of granules along its lower border, but is otherwise smooth and polished : the *carpus* has a few coarse spinules on its outer surface : the *hand*, though distinctly trigonal, is long and slender, but is enlarged at the far end; its inner and outer borders are irregularly and unequally laciniated, the teeth becoming larger and closer set towards the far end; except for a line of beading along its lower border and an occasional spinule on its upper surface, its surfaces are smooth and polished : the movable *finger* has its broad base denticulated.

The ambulatory legs are very slender and very short—only one-

117

fifth longer than the carapace : except for a line of spinules along the
posterior (lower) border of the meropodite they are smooth.

Greatest length of carapace (male) ...   ... 16 millim.
„ breadth „   ...   ... 18 „
Length of cheliped ...   . 72 „
Off Ceylon in deep-water.

Colours in spirit: chelipeds and legs purplish white, carapace dull
slaty purple.

Sub-genus AULACOLAMBRUS, A. M.-Edw.

Aulacolambrus, A. Milne-Edwards, Miss. Sci. Mex. Crust., I. p. 147.
Aulacolambrus, Miers, ' Challenger ' Brachyura, p. 97.

Pterygostomian region traversed, from the orbit to the afferent
branchial orifice, by a deep channel, which is closed and converted into
a tube by thick fringes of hairs : the lateral epibranchial spine is of
huge size : the edges of the carapace chelipeds and legs are more or
less conspicuously hairy.

Key to the Indian species of the sub-genus Aulacolambrus.

I. Carapace as long as broad, with a projecting rostrum and a
distinct post-ocular constriction; its surface closely
covered with rasp-like tubercles : carapace and legs not
conspicuously hairy.. *L. sculptus.*

II. Carapace broader than long, its surface irregularly tuberculate; rostrum not or hardly projecting: no post-ocular constriction: margins of carapace, chelipeds and legs fringed with remarkably long tangled hairs.

1. Antero-lateral border with large spines in front of the large lateral epibranchial spines : spines of inner edge of hand strongly curved upwards and outwards.. *L. curvispinis.*

2. Antero-lateral border with small teeth in front of the large lateral epibranchial spines : spines of inner edge of hand not curved.

a. No spines in middle line of carapace, or on branchial regions....... *L. hoplonotus.*

b. Some spines in middle line of carapace, and on branchial regions: spines on outer edge of hand very long.................. *L. whitei.*

Lambrus (Aulacolambrus) sculptus, A. M.-Edw.

Lambrus sculptus, A. Milne-Edwards, Nouv. Archiv. du Mus., VIII. 1872,
p. 258, pl. xiv. fig. 3.
Lambrus sculptus, Miers, ' Challenger ' Brachyura, p. 98.
Lambrus sculptus, J. R. Henderson, Trans. Linn. Soc., Zool. (2) V. 1893, p. 350.

The carapace is triangular, broad behind, and as long as broad.
The rostrum is triangular, dorsally grooved and declivous, and tapers

to a rounded point. The regions are elevated, and the median are separated from the branchial by deep furrows: all the regions are closely covered by rasp-like tubercles.

The lateral borders are tubercular, and end posteriorly in a large spine directed outwards and somewhat backwards.

Internal to this large spine is a much smaller spine; and the posterior border is tuberculate.

The chelipeds are a little more than twice the length of the carapace, with the inner and outer borders serrated, and the upper surface covered with tubercles like those on the carapace : amid the serrations five large teeth on the outer border of the hand are very conspicuous.

The ambulatory legs are slender and smooth.

The epistome is sculptured, and is very deeply excavated in the middle line.

The pterygostomian region is traversed by a canal running parallel with the buccal frame : the canal is perfectly smooth, and is closed below, and thus converted into a tube, by thick fringes of long hairs.

I believe, with Ortmann, that this species is very probably identical with *L. pisoides*, Adams and White ('Samarang' Crustacea, p. 28, pl. v. fig. 4), and perhaps with *L. diacanthus* de Haan (Faun. Japon. Crust., p. 92, pl. xxiii. fig. 1).

It is a fairly common species at the Andamans and Nicobars.

Lambrus (Aulacolambrus) hoplonotus, Ad. & Wh.

Lambrus hoplonotus, Adams and White, 'Samarang' Crust., p. 35, pl. vii. fig. 3.
Lambrus hoplonotus, A. Milne-Edwards, Nouv. Archiv. du Mus., VIII. 1872, p. 258.
Lambrus hoplonotus, Miers, Ann. Mag. Nat. Hist., 1879, Vol. IV. p. 22; and 'Challenger' Brachyura, p. 98.
Lambrus hoplonotus, Haswell, P. L. S., N. S. Wales, Vol. IV. 1879, p. 450; and Cat. Austral. Crust., p. 33.

Carapace with the outline in front of the huge lateral epibranchial spine almost semi-circular, the rostrum being extremely short and not breaking through the general outline. The carapace is granular, and has the regions well-defined but not elevated.

The symmetrically rounded antero-lateral margin is regularly festooned with little round teeth of uniform size, and ends at a great projecting lateral epibranchial spine : behind and internal to this spine is another small spine : the posterior border is finely granular. The chelipeds, legs, and margins of the carapace are fringed with long hairs ; and the pterygostomian region is channelled just as in *L. sculptus*.

The chelipeds in the male are a little more, and in the female a

119

little less than three times the length of the carapace : the arms and hands are depressed trigonal, and the fingers small : the *arm* has its inner edge sharply tuberculate, its outer edge strongly 4 or 5-spinate, its lower edge beaded, its upper surface with a row of 4 or 5 large granules : the *wrist* has three strong spines along its outer edge : the *hand* has its inner edge sharply 9 to 11-dentate, its outer edge very strongly 6 to 8-spinate, with small spinules alternating with the large spines, and its lower edge sharply and finely beaded. The ambulatory legs are perfectly smooth.

All our specimens are typical according to Adam and White's figure. This species is common at the Andamans.

Lambrus (*Aulacolambrus*) curvispinis, Miers.

Lambrus curvispinis, Miers, Ann. Mag. Nat. Hist., Vol. IV. 1879, p. 24; and 'Challenger' Brachyura, p. 98.

This species, which Miers in his latest notice of it considers to be one of the numerous varieties of *L. hoplonotus*, resembles the latter species in every particular except (1) that the rostrum ends in a little bacillar spinule ; (2) that the antero-lateral borders of the carapace instead of being crenate are powerfully spinate ; (3) that the spines along the inner edge of the palm are strongly hooked upwards and outwards ; and (4) that the inner surface of the arm bears a row of spinules.

This species, or variety, which is twice the size of *L. hoplonotus*, is also very common at the Andamans.

Lambrus (*Aulacolambrus*) whitei, A. M.-Edw.

Lambrus carinatus, Adams and White (*nec* Edw.), 'Samarang' Crust., p. 27, pl. v. fig. 3.
Lambrus whitei, A. Milne-Edwards, Nouv. Archiv. du Mus., VIII. 1872, p. 260; and Miss. Sci. Mex. Crust., I. p. 147 (foot-notes).
Lambrus whitei, Miers, 'Challenger' Brachyura, p. 98.

In the form of the carapace, the hairiness of the edges of the legs and carapace, and in the presence of the pterygostomian canal, this species almost exactly resembles the two preceding species.

The antero-lateral borders are sharply crenulate and end at a large outwardly and backwardly directed spine, internal to which is another largish spine ; while on the posterior border are four largish spines. The carapace is granular, and in the middle line are two conical spines, one on the gastric the other on the cardiac region, while on either branchial region are two similar spines.

The spinature of the chelipeds is, in disposition, similar to that

120

of *L. hoplonotus*, but the spines, especially those on the outer edge of the hand, are very much longer, slenderer, and more acute.

Several specimens, including ovigerous females, of this small species are in the Museum collection, from Arakan ; and from off Ceylon, 34 fathoms.

The figure in Adams and White is an admirable illustration of this species.

Sub-genus PARTHENOLAMBRUS, A. M.-Edw.

Parthenolambrus, A. Milne-Edwards, Miss. Sci. Mex. Crust., I. p. 148.
Parthenopoides, Miers, Journ. Linn. Soc., Zool., Vol. XIV. 1879, p. 672.
Parthenolambrus, Miers, ' Challenger ' Brachyura, p. 99.

Carapace semi-elliptical or semi-circular, with a nearly straight posterior margin, the postero-lateral angles being strongly produced. Chelipeds of no great length, never sharply serrate, and with the arms and hands indefinitely contorted. The rostrum is more or less deflexed.

Key to the Indian species of the sub-genus Parthenolambrus.

I. Carapace with the hepatic regions very prominent in the antero-lateral margin :—

    1. Carapace broader than long, strongly convex, no-dular and eroded: chelipeds less than twice the length of the carapace .... *L. tarpeius.*

    2. Carapace as long as broad, compressed, with crist-iform edges, its surface almost devoid of gra-nules: chelipeds more than twice the length of the carapace .... *L. harpax.*

II. Carapace with the hepatic regions distinct, but not marked-ly prominent :—

    1. Rostrum almost vertically deflexed : ambulatory legs *dentate*, but without true spines .... *L. calappoides.*

    2. Rostrum moderately deflexed, with a prominent median lobe : meropodites of ambulatory legs each with three rows of close sharp spines...... *L. beaumontii.*

Lambrus (Parthenolambrus) calappoides, Ad. and Wh.

Parthenope calappoides, Adams and White, ' Samarang' Crustacea, p. 34, pl. v. fig. 5.
Lambrus calappoides, Haswell, P. L. S., N. S. Wales, Vol. IV. 1879, p. 452; and Cat. Austral. Crust., p. 35.
Lambrus calappoides, Miers, Zoology of H. M. S. ' Alert,' pp. 517 and 527; and ' Challenger ' Brachyura, p. 101.
Parthenolambrus calappoides, R. I. Pocock, Ann. Mag. Nat. Hist., 1890, Vol. V. p. 75.

Carapace almost semi-circular in outline, with an indentation

121

behind the hepatic regions: the regions are well-delimited, but not carinated or sharply raised; and the surface is granular without any very large spines or nodules. The rostrum is deflexed almost vertically. The eyes are sunk in deep orbits with swollen margins. The antero-lateral margins, and sometimes the postero-lateral, are closely festooned or incised, but in an irregular manner.

On either side of the gastric region is a deep hollow; and on either side of the front part of the cardiac region is a deep foramen.

The chelipeds in the male are not twice the length of the carapace: the arm is coarsely spinate along its convex inner border, and the hand still more coarsely and bluntly spinate along its contorted upper border.

Ambulatory legs compressed, the 3rd to 5th joints having the edges irregularly dentate, this being most marked in the case of the last pair.

The animal as a whole has a sort of boiled appearance.

The species is very variable, and owing to frequent and extensive incrustation with barnacles, foraminifera, etc., is very hard to describe.

In the Museum collection are specimens from the Andamans, Mergui, Arakan, Ceylon, and Malabar coast.

Lambrus (Parthenolambrus) beaumontii, n. sp.

Very near to *Parthenope bouvieri* and *trigona*, A. M.-Edw., (*v.* Rev. et. Mag. Zool. (2) XXI. 1869, pp. 350–353).

This species comes from deepish water, and is small and very variable — the adult female, especially, being so unlike the male, that if it were found apart, it would be considered distinct.

The carapace is semicircular, the curve being broken (1) by the hepatic regions, and (2) by the projecting middle lobe of the rostrum. The elegantly curved antero-lateral borders are closely festooned by a row of thin, sharp, laciniated teeth, the bases of which are fused together; of these teeth the first three, situated on the hepatic region, are smaller than the others, which are of equal size, except the last, and this forms the summit of the salient upcurved postero-lateral angle. The postero-lateral borders are irregularly serrated, and there is a spinule in the middle of the posterior border. The regions of the carapace are very salient and form three cariniform elevations: there is usually, but not always, in the male, and seldom in the female, a recurved spinule on the gastric region, in the middle line; and generally in the male, but seldom in the female, the conical cardiac region is surmounted by one or two spinules.

The rostrum is trilobed, the small lateral lobes being formed each of a group of granules, and the larger, projecting, median lobe being spathulate, smooth, and somewhat deflexed.

The surface of the carapace is somewhat granular and eroded, but this is often concealed by a glazing of stony algæ.

The orbits have the edges finely and evenly serrate. The third joint of the antennal peduncle is spiniferous.

The segments of the sternum, as also the abdominal terga, are all deeply cut, and their surface, like that of the external maxillipeds and pterygostomian regions, is very sharply, closely and evenly granular.

The chelipeds in the male are 2⅔ times the length of the carapace; in the female hardly twice that length: in both sexes they are top-heavy, owing to the distal enlargement of the palm and the great size of the fingers; they are everywhere granular, but most markedly so on the under surface: the inner border of the arm and palm, and the upper border of the movable finger, are irregularly spinulate, the outer border of the hand may have two or three irregularly disposed blunt teeth, and that of the arm a few spicules. The ambulatory legs characterize this species, for the meropodites, in all, are compressed-trigonal with all three edges strongly, sharply and closely spinate; the anterior, and often also the posterior, margins of the next two joints also are spinate or dentate.

	Male.	Female.
Greatest length of carapace	10·5 millim.	9 millim.
„ breadth „	10·5 „	9 „
Length of chelipeds	29 „	15·5 „

Loc. Off Ceylon 32–34 fms., and off the Andamans, 41 fms.

Lambrus (Parthenolambrus) tarpeius, Ad. and Wh.

Lambrus tarpeius, Adams and White, ' Samarang' Crust., p. 35, pl. vii. fig. 2.
Lambrus tarpeius, Miers, ' Challenger' Brachyura, p. 99.

Carapace covered with numerous large nodules, and with the division into three lobes—a median and two lateral—well-marked. The hepatic region not only projects very strongly forwards, but is brought into greater prominence by the fact that the carapace is somewhat contracted behind the eyes, and excavated and constricted behind the hepatic regions themselves: the antero-lateral margins are crenulate; the produced postero-lateral angle ends in a rounded lobe-like spine, and the posterior and postero-lateral margins are irregularly and bluntly toothed.

The rostrum, which is deeply excavated and considerably deflexed, ends in a blunt point.

123

The chelipeds are massive and nodular, but even in the male are only about half as long again as the carapace.

The ambulatory legs have the 3rd, 4th and 5th joints compressed and irregularly dentate along one or both edges.

Our specimens, which are rather damaged, come from the Andamans to 20 fathoms, and from off Colombo, 26½ fathoms.

Lambrus (Parthenolambrus) harpax, Ad. and Wh.

Lambrus harpax, Adams and White, ' Samarang ' Crustacea, p. 25, pl. vi. fig. 3.
Lambrus harpax, Haswell, P. L. S., N. S. Wales, Vol. IV. 1879, p. 450; and Cat. Austral. Crust., p. 32.
Lambrus harpax, Miers, Zoology H. M. S. ' Alert,' pp. 182 and 202; and ' Challenger ' Brachyura, p. 99.

Male. Carapace depressed semi-elliptical, as long as broad, its surface almost smooth. The median region is carinated, the carina bifurcating anteriorly to enclose an elongate-triangular depression behind the eyes, and carrying a large spine in the gastric region (at the point of bifurcation), another large spine in the cardiac region, and a much smaller spine in front of the latter.

The lateral margins are cristiform, with a series of crenations and sutures indicating fused teeth ; and the hepatic region is prominent, with a cristiform edge: the postero-lateral angle is surmounted by an upturned laciniated tooth, the postero-lateral margins are dentate, and on the posterior border is a triangular tooth with an obscurely tri-lobed tip : from the bluntly laciniated tooth of the postero-lateral angle a carina runs obliquely forwards and inwards onto the posterior part of the branchial region.

The rostrum is strongly deflexed, and ends in an obscurely and unevenly trilobed tip. The chelipeds in the male are nearly 2½ times the length of the carapace, and are thin and compressed, with sharp, almost cristiform, edges : in the *arm* both the inner and outer edges are unevenly dentate, and the lower edge faintly granular: the *carpus* has the outer edge compressed and crenulate : the thin *hand* has its inner edge crenulate, has a curved line of granules on its inner surface, and some granules on its outer surface : the movable *finger* has its upper edge crenulated at base. The ambulatory legs are compressed, with the 3rd, 4th and 5th joints cristated above, especially in the last two pairs : in the last pair these joints have both margins rather strongly dentated.

Our specimen is from the Andamans.

Miers (Zoology H. M. S. ' Alert,' p. 202) considers *L. sandrockii,*

Haswell (P. L. S., N. S. Wales, Vol. IV. 1879, p. 452, pl. xxvii. fig. 2) to be identical with this species.

PARTHENOPE, Fabr.

Parthenope, Milne-Edw., Hist. Nat. Crust., I. 359, (*v. synon.*)
Parthenope, Miers, Journ. Linn. Soc., Zool., Vol. XIV. 1879, p. 668.

The form and structure of the carapace is somewhat similar to that of *Parthenolambrus*; but the genus is distinguished from *Lambrus* by the nature of the so-called basal antennal joint, which is relatively long, and nearly reaches to the level of the inferior orbital hiatus : the fingers also are much less turned inwards.

Key to the Indian species of the genus Parthenope.

I.   Carapace remarkably rugose or spinose : chelipeds nearly of the ordinary *Lambrus* form, and beset with huge spines : ambulatory legs strongly spinate :—

    1. Carapace and chelipeds beset with coarse tuber-cles and spines : carapace about ¾ as long as broad.. *P. horrida.*

    2. Carapace and chelipeds beset with spines, which are sharp and laciniate on the chelipeds : cara-pace only ⅔ as long as broad .... *P. spinosissima.*

II. The whole body and all the appendages beset with delicate paxilliform tubercles which unite to form a lace-work or frosting : chelipeds tapering, with long slender spiny fingers, nearly as long as the palm (sub-genus *Partheno-merus*).. *P. efflorescens.*

Parthenope horrida, Fabr.

Rumph, Amboin. Rariteitk. ix. 1.
? Seba, III. xix. 6-7.
- *Cancer horridus*, Linn. Syst. Nat. II. 1047, 43.
? *Cancer horridus*, Herbst, I. ii. 222, tab. xiv. fig. 88.
Parthenope horrida, Fabr., Suppl., 353.
Parthenope horrida, Leach, Zool. Misc., II. 107.
Parthenope horrida, Desmarest, Consid. Crust., p. 143, pl. xx. fig. 1.
[*Parthenope horrida*, Guérin, Icon. R. A., pl. vii. fig. 1.]
Parthenope horrida, Milne-Edwards, Hist. Nat. Crust., I. 360.
Parthenope horrida, Cuv. Regn. An., pl. xxvi. fig. 2.
Parthenope horrida, A. Milne-Edwards, Nouv. Archiv. du Mus., VIII. 1872, p. 255.
Parthenope horrida, Martens, Archiv. fur Naturges., XXXVIII. 1872, p. 86 (note on habitat).
Parthenope horrida, Miers, Phil. Trans., Vol. 168, p. 486.
Parthenope horrida, Nauck, Z. Wiss. Zool., XXIV. 1880, p. 44 (gastric teeth).
Parthenope horrida, C. W. S. Aurivillius, Kongl. Sv. Vet. Ak., Handl. XXIII. No. 4, 1888-89, p. 60.
[*Parthenope horrida*, F. Muller, Verh. Ges., Basel., VIII. p. 473].

125

Carapace somewhat pentagonal ; its length not quite ¾ its breadth ; its surface deeply eroded, strongly rugose, and sharply tubercular : its postero-lateral angle much produced outwards : antero-lateral margin coarsely spinate : postero-lateral and posterior margins granular, the former with a coarse spine. Rostrum short, moderately deflexed, ending in a blunt inter-antennulary tooth. Orbits circular, deep.

Chelipeds huge, one much larger than the other, the larger twice the length of the carapace (in the female), covered with large coarse granular spines.

Ambulatory legs stout, spiniferous ; the dactylus smooth : the meropodite, in all, is compressed-trigonal, with all the edges spinate.

The under surface of the body has a worm-eaten appearance : the sternum is deeply pitted, with a deep crescentic excavation between the chelipeds.

The abdomen (of the female) with a series of deep excavations along either side.

Off Ceylon, 34 fathoms.

Parthenope spinosissima, A. M.-Edw.

Seba, III. xxii. 2 and 3.
Parthenope spinosissima, A. M.-Edw., in Maillard's l'île Réunion, Annexe F, p. 8, pl. xviii.
Parthenope spinosissima, Alcock, J. A. S. B., 1893, Pt. ii. p. 177.

Carapace in the form of an equilateral triangle, its length only about ⅔ its breadth ; its surface strongly rugose, and sharply tubercular and spinate : the antero-lateral borders are armed with large laciniate spines ; the posterior and postero-lateral borders are sharply spinate : the strongly-produced and spinate postero-lateral angle runs forwards as a carina onto the branchial regions.

The three lobes of the gastric region are greatly inflated.

The rostrum is vertically deflexed, and ends in a strong sharp inter-antennulary spine.

The chelipeds are very little asymmetrical, and are beset, nearly up to the tips of the fingers, with great ramose and laciniate spines.

The ambulatory legs are armed with extremely sharp teeth almost up to the tip of the dactylus.

The abdomen of the female has a median double series, and on either side a single series, of sharp spines.

A male and female from the Bay of Bengal, 88 fathoms.

Sub-genus PARTHENOMERUS, *nov.*

Characterized by the chelipeds, which have a thigh-shaped meropodite, and taper to the fingers, which are nearly as long as the palm, and are extremely slender.

Parthenope (*Parthenomerus*) *efflorescens,* n. sp.

Carapace triangular, not quite ¾ as long as broad; its entire surface, above and below, as also that of the sternum, of the abdomen (in the female), and of all the exposed appendages—from the eye-stalks to the last pair of ambulatory legs, covered with a lace-work, or frosting, formed by the partial contact of very delicate crisply paxilliform granules. There are no large tubercles, and, except on the arm hand and fingers, no spines. On the arm, namely, there are two or three teeth with acicular tips, on both the lower-inner, and the upper-inner borders; on the hand there are three needle-like teeth on the upper-inner, and three on the lower-inner borders; and the fingers are everywhere beset with long needle-like spines. The rostrum is nearly vertically deflexed.

Only one cheliped remains in our unique specimen; and it, which is a little over twice the length of the carapace, has a most curious tapering form: the meropodite is huge and thigh-shaped, decreasing in size distally; the carpus is slenderer than the end of the meropodite; and the hand is still slenderer than the carpus: the fingers are long— nearly as long as the palm—are extremely slender, and, as already noted, are beset with long slender spines.

A single female, from the Andaman Sea, 36 fathoms.

CRYPTOPODIA, Edw.

Cryptopodia, Milne-Edwards, Hist. Nat. Crust., I. 360.
Cryptopodia, Miers. Journ. Linn. Soc. (Zool.), XIV. p. 669.
Cryptopodia, Miers, 'Challenger' Brachyura, p. 101.

Carapace very broadly triangular, with very large lateral clypeiform vaulted expansions which completely conceal the ambulatory legs, and are prolonged posteriorly far beyond the base of the abdomen; a large space between the gastric and the cardiac regions is triangular and concave. The rostrum is nearly horizontal, spatuliform and very prominent. The pterygostomian regions are smooth, not ridged. The orbits are very small, nearly circular, with a suture in the superior margin. The epistome is well developed; the antennulary fossæ are narrow and somewhat oblique. The abdomen, in the male, is five-jointed; the third to fifth segments coalescent. The eyes are very small and retractile. The basal antennal joint is slightly dilated and does not nearly reach the internal orbital hiatus, which is filled by the second joint. The buccal cavity and external maxillipeds are small. The ischium-joint of the maxillipeds is not produced at its antero-internal angle; the merus is distally truncated, with the antero-external angle slightly produced, the interior margin notched below the antero-internal angle. The chelipeds are nearly as in *Lambrus;* the merus-joint has a wing-like lobe on the posterior margin near to the distal extremity; the

palms of the chelipeds are elongated, tricarinated, and dentated (as in
Lambrus); fingers short. The ambulatory legs are slender, decrease
successively but slightly in length, and have the fourth, fifth and sixth
joints more or less distinctly carinated ; dactyli nearly straight.

Cryptopodia fornicata, (Fabr.)

Cancer fornicatus, Fabr., Ent. Syst., II. 453.
Cancer fornicatus, Herbst, I. ii. 204, pl. xiii. figs. 79–80.
Parthenope fornicata, Fabr., Suppl., p. 352.
Maia fornicata, Latr., Hist. Nat. Crust., VI. 104.
Oethra fornicata, Desmarest, Consid. Crust., p. 110.
Cryptopodia fornicata, Milne-Edwards, Hist. Nat. Crust., I. 362 (*v. synon.*)
Cryptopodia fornicata, de Haan, Faun. Japon. Crust., p. 90, pl. xx. figs. 2 and 2*a*;
and (?) Adams and White, 'Samarang' Crust., p. 32, pl. vi. fig. 4; and Dana, U. S.
Expl. Exp. Crust., pt. I. p. 140; and Stimpson, Proc. Ac. Nat. Sci., Philad., 1857.
p. 220; and Haswell, P. L. S., N. S. Wales, Vol. IV. 1879, p. 454; and Cat. Austral.
Crust., p. 37; and E. Nauck, Z. Wiss. Zool., 1880 (gastric teeth); and Miers, Zool.
H.M.S. 'Alert,' pp. 182 and 203; and 'Challenger' Brachyura, p. 102; and A. O.
Walker, Journ. Linn. Soc., Zool., Vol. XX. 1890, p. 109; and J. R. Henderson, Trans.
Linn. Soc., Zool., (2) V. 1893, p. 351.

Carapace broadly triangular, depressed : the antero-lateral margins
more or less laciniated, the posterior and postero-lateral margins
forming one strong curve, the edge of which is either unbroken or
shows very faint traces of crenulation : the surface of the carapace is
in the main smooth, but the triangular depression is a little pitted and
is bounded by lines of granules, the lateral lines being produced well
across the branchial regions. The rostrum is prominent, blunt-pointed,
about as long as broad, and has its edge very faintly crenulate.

The chelipeds are considerably less than twice the length of the
carapace, and have massive sharply trigonal joints, with most of the
edges strongly cristiform; and the fingers are massive and strongly
incurved as in *Lambrus* : in the arm, the cristiform inner and outer
edges are sharply laciniate, the latter being strongly alate, while the
lower edge is beaded : in the *carpus* the outer edge only is cristiform :
in the hand both the inner and outer edges are strongly cristiform and
laciniate, the lower edge being crenate.

The ambulatory legs have both edges of the merus raised into
spiniform crests, and the upper edges of the next two joints carinate.

In the Museum collection are numerous specimens from Palk
Straits, Andamans and Persian Gulf.

Cryptopodia angulata, Edw. and Lucas.

Cryptopodia angulata, Edw. and Lucas, Archiv. du Mus., Vol. II. 1841, p. 481,
pl. xxviii. figs. 16–19.

Carapace convex, sharply pentagonal, with all the edges deeply dentated, and all the angles produced to form curved spines; in addition there is a second spine in front of the spine of either antero-lateral angle, and the part of the posterior border that is co-extensive with the abdomen is demarcated on either side by a strong spine. The rostrum ends in a sharp point. The triangular depression of the carapace is very deep, and the lines which bound it are granular; there is an irregular patch of granules on either branchial region, and there is a line of granules passing forwards from the apex of the triangular depression to the base of the rostrum on either side.

The chelipeds are much as in *C. fornicata*, with the exception that the carpus is semi-globular, and that the inner and outer margins both of the hand and arm are armed with sharp laciniate spines. The ambulatory legs have the merus simply carinate above, spinate-carinate below, the carpus and propodite carinate, and the dactylus strongly carinate on both edges so as to form a swimming blade.

Orissa coast, 20–25 fathoms. Malabar coast, 28 fathoms.

In a large male from the Malabar coast, the carapace is much more granular; and the chelipeds have the spinature much more acute and laciniate, and their surfaces—especially the under surface—granular instead of nearly smooth.

Cryptopodia angulata, var. *cippifer*, nov.

In this variety the only differences are: (1) that the semi-globular carpus has a few granules on its upper surface; and (2) that the triangular hollow in the middle of the carapace is rather deeper, and has certain large erect definitely-placed spines on the ridges that bound the hollow, namely,—two close together side by side in the middle line, in front; one at either branchial angle; and one in the middle line posteriorly, on the summit of the cardiac region.

These spines are present in six specimens of both sexes, but are most pronounced in the male.

Loc. Karáchi.

The largest specimen, female, has an extreme breadth of carapace of 45 millim.

HETEROCRYPTA, Stimpson.

Heterocrypta, Stimpson, Ann. Lyc. Nat. Hist., New York, Vol. X. 1874, p. 102.
Heterocrypta, A. Milne-Edwards, Miss. Sci. Mex., Crust., I. p 166.
Heterocrypta, Miers, J. L. S., Zool., Vol. XIV. 1879, p. 669; and 'Challenger' Brachyura, p. 102.

129

Differs from *Cryptopodia* in the following characters :—

The posterior border of the carapace slightly overlaps the abdomen, but is not distinctly produced ; the lateral clypeiform expansions are also less produced, so that the legs when even moderately extended can be seen beyond them.

The pterygostomian and sub-hepatic regions are traversed by a granular ridge which runs parallel to the antero-lateral border from the angle of the buccal cavity to the base of the chelipeds.

Heterocrypta investigatoris, n. sp.

Carapace broadly pentagonal ; the posterior border almost straight, and crenulated ; the other borders sharply dentate. The central depression of the carapace is semi-circular and very deep, with the boundary raised into a carina : the horns of the semi-circle end each in a boss or mammillary tubercle, from which a carina runs backwards to the posterior angle of the carapace. The rostrum is very large and prominent, shaped like a leaf : its surface is smooth : that of the carapace is either smooth or granular—the granules, when present, being most abundant on the posterior part of the branchial regions.

The chelipeds, which are twice the length of the carapace, have both the inner and outer edges of the *arm* sharply dentate (but not alate as in *Cryptopodia*), and the lower edge beaded : the *carpus* is subglobular : the *hand* has both the inner and the outer edges bluntly dentate, and the under surface closely covered with bead-like granules.

The ambulatory legs have the upper edges of the 3rd, 4th, and 5th joints sharply carinate : the meropodite also, in the case of the first two pairs of legs, has a single row of teeth or spines along its lower edge, and in the case of the last two pairs of legs has a double row of spines along the lower edge.

Like all the species of this genus, this species is small, the breadth of the carapace in the largest specimen being 18 millim.

It is not uncommon off rocky parts of the coasts of India up to and about 30 fathoms. It would seem to be allied to the *Cryptopodia contracta* of Stimpson (Proc. Ac. Nat. Sci., Philad., 1857, p. 220).

OETHRA, Leach.

Oethra, Leach.
Oethra, Milne-Edwards, Hist. Nat. Crust., I. 370.
Oethra, A. Milne-Edwards, Miss. Sci. Mex., Crust., I. p. 170 (*v. synon.*).
Oethra, Miers, Journ. Linn. Soc., Zool., Vol. XIV. 1879, p. 669.

The carapace is regularly oval (transversely), with its surface strongly rugose, and its antero-lateral edges somewhat upturned. The rostrum is obsolete, not breaking the general oval outline. The eyes are small; and the orbits are nearly circular, with two sutures in the upper border, and a hiatus at the inner inferior angle, which is filled by the second joint of the antennary peduncle.

The antennulary fossæ are squarish, and are nearly filled by the large angular basal joint, internal to which the rest of the antennule folds obliquely.

The basal antennal joint is oblong and angular, and reaches to the internal orbital canthus: the antennary flagella are rudimentary.

The external maxillipeds completely close the buccal frame: their inner border is extremely straight and sharp cut: their palp is inserted at the antero-internal angle of the merus, and folds out of sight.

The chelipeds are about equal in length to the carapace: they have somewhat the *Lambrus* form—having sharply prismatic joints and large inturned fingers, but are concave on the upper surface.

The ambulatory legs are short, and decrease gradually in length : they are all strongly dentate-carinate, or cristate.

The abdomen of the female (and young male) consists of seven segments.

Oethra scruposa, L.

[*Cancer scruposus*, Linn., Mus. Lud. Ulr., p. 450.]
Cancer polynome, Herbst, III. ii. 23, tab. liii. figs. 4-5.
[*Oethra depressa*, Lamk., Hist. Anim. Sans. Vert., V. 265.]
Oethra depressa, Desmarest, Consid. Crust., p. 110, pl. x. fig. 2.
[*Oethra depressa*, Guérin, Icon. R. A., pl. xii. fig. 3.]
Oethra scruposa, Milne-Edwards, Hist. Nat. Crust., I. 371.
Oethra scruposa, Cuv., R. A., pl. xxxviii. fig. 2.
Oethra scruposa, Stimpson, Proc. Ac. Nat. Sci., Philad., 1857, p. 221.
Oethra scruposa, A. M.-Edw., in Maillard's l'île Réunion, Annexe F., p. 3; and Nouv. Archiv. du Mus., VIII. 1872, p. 263.
Oethra scruposa, Henderson, Trans. Linn. Soc., Zool., (2) V. 1893, p. 351.
[*Oethra scruposa*, F. Muller, Verh. Ges., Basel, VIII. 473.]

(*Oethra scruposa*, var. *scutata* A. Milne-Edwards, Miss. Sci. Mex., Crust., I. p. 170, pl. xxxi. fig. 2=*Oethra scutata*, S. I. Smith, Amer. Journ. Sci., etc., XLVIII. 1869, p. 120; and Ann. Mag. Nat. Hist., 1869, Vol. IV. p. 230, is considered by M. A. Milne-Edwards to be only a variety of the Linnæan type.)

The antero-lateral borders are divided into 6 or 7 indistinct lobes by deep narrow sutures, each fold being again subdivided near the edge by a faint crest.

The gastric region is extremely prominent, and is divided into two lobes by a broad longitudinal channel, each lobe being sparsely granular : the branchial regions are also somewhat convex near their middle, the

131

convexities being granular : the rest of the carapace is somewhat concave.

The chelipeds and ambulatory legs are rough : the chelipeds have
the lower edge sharply dentate, and the outer edge of the carpus sharply dentate : the ambulatory legs have the 3rd, 4th and 5th joints carinate or cristate above, and the 3rd and 5th joints cristate below : the
dactyli are cristate on both edges, and end in little claws.

The abdomen is deeply sculptured.

In the Museum collection is a male from the Andamans, and a
female from Ceylon.

Sub-family II. EUMEDONINÆ, Miers.

Miers, Journ. Linn. Soc., Zool., Vol. XIV. 1879, p. 670.

Carapace rhomboidal or pentagonal, with a spine at the junction of
the antero-lateral and postero-lateral borders. Rostrum usually bifid
or emarginate. Surface of carapace nearly flat. Chelipeds of moderate size and length.

Key to the Indian genera of the sub-family EUMEDONINÆ.

I. Floor of the orbit not in contact with the front, but leaving
   a hiatus which is more or less filled by the second joint
   of the antennal peduncle. Chelipeds armed with large
   spines : ambulatory legs compressed :—

     1. Spine of antero-lateral angle of carapace direct-
       ed forwards.. ZEBRIDA.

     2. Spine of antero-lateral angle directed straight
       outwards; last pair of legs dorsal in position... EUMEDONUS.

II. Floor of the orbit meeting the front, so as to completely
    exclude the antennal peduncle from the orbit: chelipeds
    not armed : ambulatory legs not compressed.................. CERATOCARCINUS.

ZEBRIDA, Adams and White.

Zebrida, Adams and White, ' Samarang ' Crustacea, p. 23.
Zebrida, Miers, J. L. S., Zool., Vol. XIV. 1879, p. 670.

Carapace sub-rhomboidal, flattened, with the rostrum formed by
two large, acute, laminar, almost parallel teeth ; and with the antero-
lateral angles produced to form two similar laminar teeth projecting
forwards in a plane parallel to the rostrum.

Orbits circular, their inner canthus being filled by part of the
antennal peduncle.

The antennules fold obliquely. The antennæ are entirely concealed
beneath the rostrum : their flagellum is well developed ; and their
basal joint is longish, reaching to the inner canthus of the orbit.

The chelipeds are stout but short, the legs are compressed, and both are armed with large laminar spines of the same type as those that form the rostrum and the antero-lateral margins of the carapace. The ambulatory legs are subchelate much as in *Acanthonyx.*

Zebrida adamsii, White.

Zebrida adamsii, White, P. Z. S., 1847, p. 121; and Ann. Mag. Nat. Hist., 1848, Vol. I. p. 223; and ' Samarang' Crustacea, p. 24, pl. vii. fig. 1.

Zebrida adamsii, J. R. Henderson, Trans. Linn. Soc., Zool., (2) V. 1893, p. 351.

Zebrida longispina, Haswell, P. L. S., N. S. Wales, Vol. IV. 1879, p. 454, pl. xxvii. fig. 3; and Cat. Austral. Crust., p. 38.

Body of a light delicate madder pink, the carapace with darker (liver-coloured) parallel longitudinal bands and alternating streaks, the legs and chelipeds with broad somewhat oblique cross-bands of the same darker colour : the median longitudinal dark band, and a band on either side of it, extend, discontinuously, from the carapace along the abdomen.

The entire integument of the body and limbs is smooth, hard, and polished. The chelipeds are stout, with short squat joints : the *arm* is trigonal with sharp-cut laminar edges, the upper and lower of which end in sharp teeth; its broad distal end is also dentate : the *wrist* is surmounted by three laminar teeth disposed in a triangle : the *hand* has its upper edge raised into a compressed tooth.

Of the ambulatory legs the 3rd, 4th, and 5th joints are strongly compressed, with the upper edges sharply and acuminately carinate ; the fifth joint is enlarged distally, and the strongly recurved dactylus is retractile against it in the manner of a subchela.

In the Museum collection are a male and female from the coast of Travancore.

EUMEDONUS, Edw.

Eumedonus, Edw., Hist. Nat. Crust., I. 349.

Eumedonus, Miers, J. L. S., Zool., Vol. XIV. 1879, p. 670.

Carapace depressed, pentagonal : rostrum large, strongly prominent, bifurcate only near the tip. Orbits circular ; their internal hiatus occupied by part of the antennal peduncle. Antennules folding obliquely ; their basal joint of large size.

Antennæ entirely concealed beneath the front ; both the peduncle and the flagellum short. Chelipeds more massive than the other legs, and in the male much longer ; armed with large spines. Ambulatory legs compressed ; their third joint cristate ; the second pair a little shorter than the third ; the fifth pair dorsal in position. The abdomen in both sexes consists of seven separate segments.

Eumedonus zebra, n. sp.

Carapace, in spirit, of a yellow colour, and traversed fore-and-aft by five broad parallel liver-coloured bands—a median and two lateral : the median and the inner lateral band on either side being continued a certain distance on to the abdomen.

The carapace is sharply pentagonal, the antero-lateral angles being sharp and directed straight outwards.

The rostrum forms a long, broad, sub-triangular lamina bifurcated near the tip.

The chelipeds in the female are about the same length as the carapace : the ischium has a sharp tooth on its inner border, the merus has one on its inner and one on its upper margin, the carpus has a very strong one on its upper border, and the hand has two on its upper border : the legs have the merus strongly compressed, with the upper border dentate or cristate, and the dactyli are strongly recurved.

Two ovigerous females from off Ceylon, 32 fms : the extreme length of the carapace of the larger specimen is 10 millim.

CERATOCARCINUS, Adams and White.

Ceratocarcinus, Adams and White, Proc. Zool. Soc., p. 57, 1847; and 'Samarang' Crust., p. 33.

Ceratocarcinus, Miers, Journ. Linn. Soc., (Zool.) XIV. p. 670, 1879; and 'Challenger' Brachyura, p. 104.

Carapace sub-hexagonal, about as broad as long, with the dorsal surface nearly flat, spinose or tuberculated. The spines of the rostrum are elongated, acute, and separated by a rather wide interspace, and there is a well-developed lateral epibranchial spine. The orbits are small and circular, and the sub-ocular lobe joins the front, so as completely to exclude the antennæ from the orbits. The basal joint of the antennæ is slender and like the greater part of these appendages is hidden beneath the front. The external maxillipeds are small, the ischium-joint not produced at its antero-internal angle, the merus distally truncated, not produced at the antero-external angle, and scarcely emarginate at the antero-internal angle, where the next joint articulates. The chelipeds are relatively slender and somewhat elongated, with the joints not dilated, the merus and carpus sometimes armed with spines ; the dactyli acute and shorter than the palms ; the ambulatory legs are slender, with the joints not dilated, the merus sometimes armed with a distal spine ; the dactyli nearly straight.

Ceratocarcinus longimanus, Ad. and Wh.

Ceratocarcinus longimanus, White, P. Z. S., 1847, p. 57; and Ann. Mag. Nat. Hist., 1847, Vol. XX. p. 62; and 'Samarang' Crustacea, p 34, pl. vi. fig. 6.

Ceratocarcinus longimanus, Miers, 'Challenger' Brachyura, p. 105.

Carapace hexagonal : the spines of the rostrum far apart : lateral angles of the carapace in the form of stout outstanding spines the tips of which are turned forwards : a pair of sharp tubercles in the middle line behind the rostrum—these being tufted with hairs.

Chelipeds stout, about twice the length of the carapace and rostrum, finely granular, and longitudinally grooved.

A single specimen of this small species, from the Malacca Straits, is in the Museum Collection.

Appendix to sub-family ACANTHONYCHINÆ.

MENÆTHIOPS, n. gen.

Closely allied to *Menæthius.*

Carapace pyriform, its surface smooth beneath a pubescent covering. The rostrum consists of two acute slender spines of moderate length, which are in the closest contact throughout.

The eyes, which are movable forwards but not retractile, are in great part concealed beneath a large, very conspicuous, laminar supraocular spine. No post-ocular spine. [A spinule is present on the ventral aspect of the hepatic region of the single species.] The basal antennal joint is broad ; and the mobile portions of the antennæ are visible, from above, on either side of the rostrum.

The external maxillipeds have the merus as broad as the ischium, and the palp inserted at the antero-internal angle of the merus.

The ambulatory legs, of which the first pair are longer than the rest, have strongly recurved prehensile dactyli.

The chelipeds in the female (male unknown) are not enlarged.

The abdominal segments in the female appear to be all distinct.

This genus has a superficial resemblance to *Oregonia*, Dana ; but in *Oregonia* there is a large post-ocular spine, quite distinct from the hepatic angle, and the eyes are said to be retractile against this spine.

Menæthiops bicornis, n. sp.

Body and legs tomentose, with additional long scattered setæ.

Carapace pyriform, somewhat *Achæus*-like in shape, there being a slight constriction behind the eyes, and another slight constriction behind the hepatic regions : the gastric and cardiac regions very prominent, the branchial regions prominent : the surface, when denuded, smooth, except for a granular ridge on the pterygostomian regions ; the hepatic regions are laterally rather prominent, and carry a small spinule

135

visible from above, on the ventral aspect of the antero-external angle, as well as a much smaller spinule on the dorsal aspect. There is also a spinule, in the middle line, on the gastric region, and one on the cardiac region, as well as one near the middle of either branchial region.

The rostrum consists of two slender acute spines, which are about one-fourth the length of the carapace proper, and are in the closest contact up to the very tips.

The eyes are movable forwards but are quite non-retractile backwards, and are in great part concealed beneath a large laminar supra-ocular spine, which has its anterior angle produced forwards and its posterior angle produced outwards. No post-ocular spine.

[The spinule on the ventral surface of the hepatic angle is in no sense a post-ocular spine.]

The basal antennal joint is broad and has its outer edge irregularly wavy, somewhat as in Dana's figure of *Oregonia gracilis* (U. S. Expl. Exp., Crust., I. pl. iii, fig. 2*b*.) ; it sharp antero-external angle is, like the following joints and the flagellum, plainly visible, from above, beside the rostrum : the mobile portion of the antenna is rather more than half the length of the carapace and rostrum.

The chelipeds in the female are not stouter than the other legs, and are shorter than the carapace and rostrum : their palm is nearly twice the length of the fingers, which meet only at the tip.

The ambulatory legs all have slender joints and a strongly recurved prehensile dactylus : the first pair, which are the longest, are, in the female, a little longer than the carapace and rostrum.

A single egg-laden female has the following dimensions :—

Length of carapace and rostrum	...	...	$6\cdot2 + 2 = 8\cdot2$ millim.
Greatest breadth of carapace	...	...	...	6·0 ,,
Length of chelipeds	...	...	...	...	7·0 ,,
Length of first ambulatory legs	...	...	...	8·5 ,,

Loc. Kárachi.

The place of the above genus in the " Key to the Indian genera of the sub-family *Acanthonychinæ* " (pp. 190 and 191 *ante*), is with *Huenia* and *Menæthius*, from both of which it is easily diagnosed (1) by the *Pisa*-like rostrum, consisting of two sharp slender spines in the closest contact throughout their extent, and (2) by the large antennary flagellum and by the eroded outer edge of the basal antennal joint. It has, indeed, the closest natural relations with *Menæthius*.

The unique specimen has only just been received along with the " Investigator " collections of the season 1894-95.

EXPLANATION OF PLATES.

PLATE III.

Fig. 1. Lambrachœus remifer, ♂.
,, 2. Physachœus ctenurus, ♂ ; 2*a*. abdomen of ♀ × 4; 2*b*. abdomen of
   ♂ × 4.
,, 3. Physachœus tonsor, ♀.
,, 4. 4*a*. Grypachœus hyalinus, ♀.

PLATE IV.

Fig. 1. 1*a*. Inachoides dolichorhynchus, ♂.
,, 2. 2*a*. Apocremnus indicus, ♂.
,, 3. Naxia investigatoris, ♂.
,, 4. Macrocœloma nummifer, ♂. ·
,, 5. Maia gibba, ♂.

PLATE V.

Fig. 1. Achœus cadelli, ♂.
,, 2. 2*a*. Chorilibinia andamanica.
,, 3. Callodes malabaricus, ♀.
,, 4. 4*a*. Paratymolus hastatus, ♀.

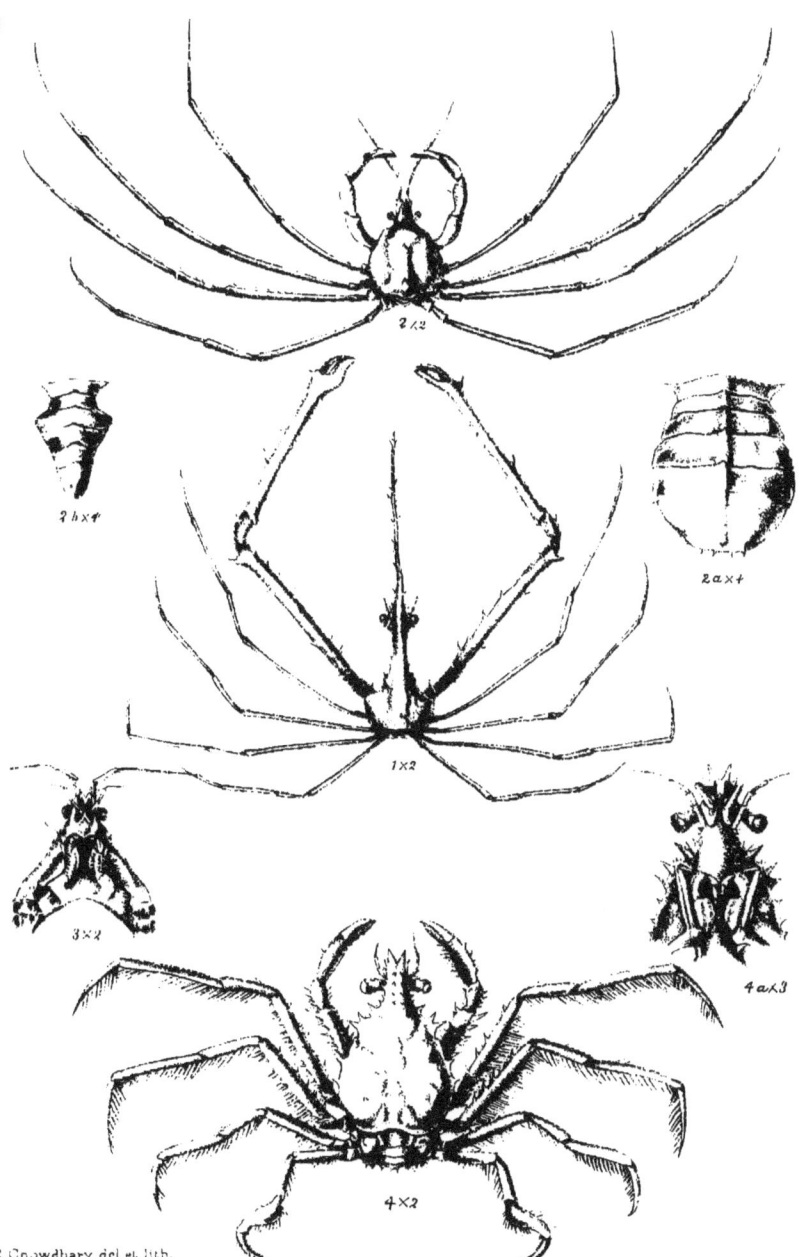

A. C. Chowdhary del et lith.

1. Lashbrachæus ramifer.   2. Physachæus ctenurus.
3. Physachæus tonsor.   4. Grypachæus   hyalinus.

1a x 3

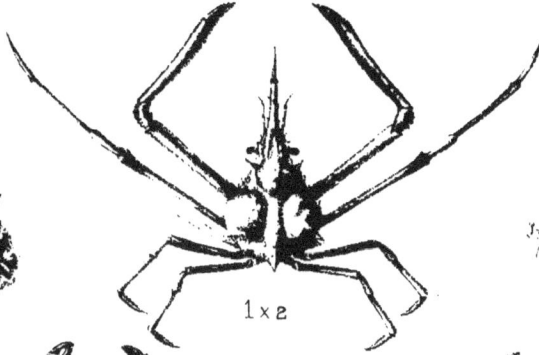

1 x 2

2a x 4

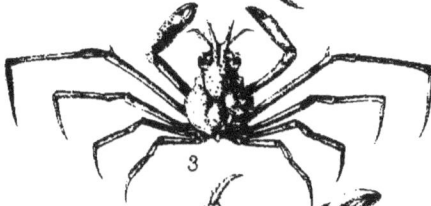

3

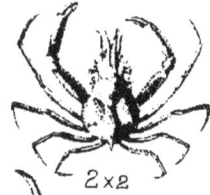

2 x 2

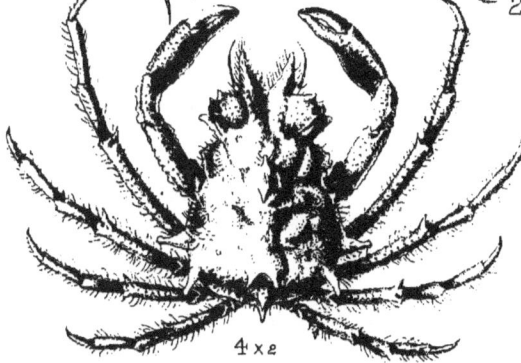

4 x 2

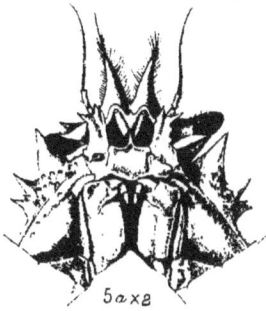

5a x 2

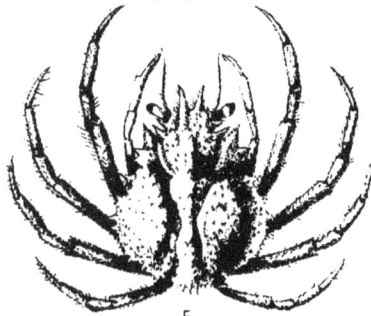

5

Chowdhary & S C Mondul del et lith

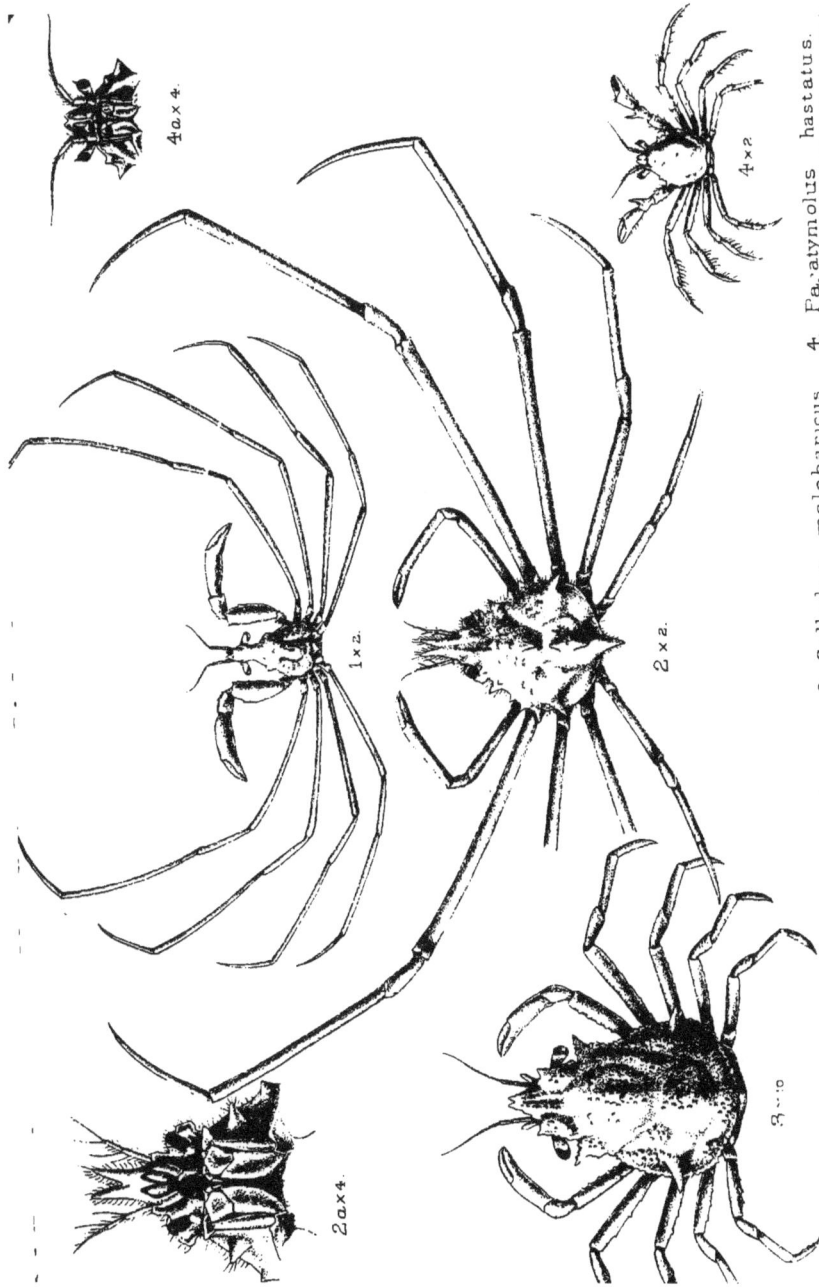

4a×4.

4×2.

Fa. atynolus hastatus.
A. Crowdney lith.

1×2.

2×2.

3. Collodes malabaricus   4. Fa. atynolus hastatus.

2a×4.

3×o.

1. Achaeus cadelli.   2. Chorilibinia andamanica.

www.ingramcontent.com/pod-product-compliance
Lightning Source LLC
Chambersburg PA
CBHW021816190326
41518CB00007B/613